A Python Data Analyst's Toolkit

Learn Python and Python-based Libraries with Applications in Data Analysis and Statistics

Gayathri Rajagopalan

Apress®

A Python Data Analyst's Toolkit: Learn Python and Python-based Libraries with Applications in Data Analysis and Statistics

Gayathri Rajagopalan

ISBN-13 (pbk): 978-1-4842-6398-3 ISBN-13 (electronic): 978-1-4842-6399-0
https://doi.org/10.1007/978-1-4842-6399-0

Managing Director, Apress Media LLC: Welmoed Spahr
Acquisitions Editor: Celestin Suresh John
Development Editor: James Markham
Coordinating Editor: Aditee Mirashi

Cover designed by eStudioCalamar

Cover image designed by Freepik (www.freepik.com)

Distributed to the book trade worldwide by Springer Science+Business Media New York, 1 New York Plaza, Suite 4600, New York, NY 10004-1562, USA. Phone 1-800-SPRINGER, fax (201) 348-4505, e-mail orders-ny@springer-sbm.com, or visit www.springeronline.com. Apress Media, LLC is a California LLC and the sole member (owner) is Springer Science + Business Media Finance Inc (SSBM Finance Inc). SSBM Finance Inc is a **Delaware** corporation.

For information on translations, please e-mail booktranslations@springernature.com; for reprint, paperback, or audio rights, please e-mail bookpermissions@springernature.com.

Apress titles may be purchased in bulk for academic, corporate, or promotional use. eBook versions and licenses are also available for most titles. For more information, reference our Print and eBook Bulk Sales web page at http://www.apress.com/bulk-sales.

Any source code or other supplementary material referenced by the author in this book is available to readers on GitHub via the book's product page, located at www.apress.com/978-1-4842-6398-3. For more detailed information, please visit http://www.apress.com/source-code.

Printed on acid-free paper

This book is dedicated to my daughter.

Table of Contents

About the Author

Gayathri Rajagopalan works for a leading Indian multinational organization, with ten years of experience in the software and information technology industry. She has degrees in computer engineering and business adminstration, and is a certified Project Management Professional (PMP). Some of her key focus areas include Python, data analytics, machine learning, statistics, and deep learning. She is proficient in Python, Java, and C/C++ programming. Her hobbies include reading, music, and teaching programming and data science to beginners.

About the Author

About the Technical Reviewer

Manohar Swamynathan is a data science practitioner and an avid programmer, with over 14 years of experience in various data science related areas that include data warehousing, Business Intelligence (BI), analytical tool development, ad hoc analysis, predictive modeling, data science product development, consulting, formulating strategy, and executing analytics programs. He's had a career covering the life cycle of data across different domains such as US mortgage banking, retail/ecommerce, insurance, and industrial IoT. He has a bachelor's degree with a specialization in physics, mathematics, and computers, and a master's degree in project management. He's currently living in Bengaluru, the Silicon Valley of India.

Acknowledgments

This book is a culmination of a year-long effort and would not have been possible without my family's support. I am indebted to them for their patience, kindness, and encouragement.

I would also like to thank my readers for investing their time and money in this book. It is my sincere hope that this book adds value to your learning experience.

Introduction

I had two main reasons for writing this book. When I first started learning data science, I could not find a centralized overview of all the important topics on this subject. A practitioner of data science needs to be proficient in at least one programming language, learn the various aspects of data preparation and visualization, and also be conversant with various aspects of statistics. The goal of this book is to provide a consolidated resource that ties these interconnected disciplines together and introduces these topics to the learner in a graded manner. Secondly, I wanted to provide material to help readers appreciate the practical aspects of the seemingly abstract concepts in data science, and also help them to be able to retain what they have learned. There is a section on case studies to demonstrate how data analysis skills can be applied to make informed decisions to solve real-world challenges. One of the highlights of this book is the inclusion of practice questions and multiple-choice questions to help readers practice and apply whatever they have learned. Most readers read a book and then forget what they have read or learned, and the addition of these exercises will help readers avoid this pitfall.

The book helps readers learn three important topics from scratch – the Python programming language, data analysis, and statistics. It is a self-contained introduction for anybody looking to start their journey with data analysis using Python, as it focuses not just on theory and concepts but on practical applications and retention of concepts. This book is meant for anybody interested in learning Python and Python-based libraries like Pandas, Numpy, Scipy, and Matplotlib for descriptive data analysis, visualization, and statistics. The broad categories of skills that readers learn from this book include programming skills, analytical skills, and problem-solving skills.

The book is broadly divided into three parts – programming with Python, data analysis and visualization, and statistics. The first part of the book comprises three chapters. It starts with an introduction to Python – the syntax, functions, conditional statements, data types, and different types of containers. Subsequently, we deal with advanced concepts like regular expressions, handling of files, and solving mathematical problems with Python. Python is covered in detail before moving on to data analysis to ensure that the readers are comfortable with the programming language before they learn how to use it for purposes of data analysis.

The second part of the book, comprising five chapters, covers the various aspects of descriptive data analysis, data wrangling and visualization, and the respective Python libraries used for each of these. There is an introductory chapter covering basic concepts and terminology in data analysis, and one chapter each on NumPy (the scientific computation library), Pandas (the data wrangling library), and the visualization libraries (Matplotlib and Seaborn). A separate chapter is devoted to case studies to help readers understand some real-world applications of data analysis. Among these case studies is one on air pollution, using data drawn from an air quality monitoring station in New Delhi, which has seen alarming levels of pollution in recent years. This case study examines the trends and patterns of major air pollutants like sulfur dioxide, nitrogen dioxide, and particulate matter for five years, and comes up with insights and recommendations that would help with designing mitigation strategies.

The third section of this book focuses on statistics, elucidating important principles in statistics that are relevant to data science. The topics covered include probability, Bayes theorem, permutations and combinations, hypothesis testing (ANOVA, chi-squared test, z-test, and t-test), and the use of various functions in the Scipy library to enable simplification of tedious calculations involved in statistics.

By the end of this book, the reader will be able to confidently write code in Python, use various Python libraries and functions for analyzing any dataset, and understand basic statistical concepts and tests. The code is presented in the form of Jupyter notebooks that can further be adapted and extended. Readers get the opportunity to test their understanding with a combination of multiple-choice and coding questions. They also get an idea about how to use the skills and knowledge they have learned to make evidence-based decisions for solving real-world problems with the help of case studies.

CHAPTER 1

Getting Familiar with Python

Python is an open source programming language created by a Dutch programmer named Guido van Rossum. Named after the British comedy group Monty Python, Python is a high-level, interpreted, open source language and is one of the most sought-after and rapidly growing programming languages in the world today. It is also the language of preference for data science and machine learning.

In this chapter, we first introduce the Jupyter notebook – a web application for running code in Python. We then cover the basic concepts in Python, including data types, operators, containers, functions, classes and file handling and exception handling, and standards for writing code and modules.

The code examples for this book have been written using Python version 3.7.3 and Anaconda version 4.7.10.

Technical requirements

Anaconda is an open source platform used widely by Python programmers and data scientists. Installing this platform installs Python, the Jupyter notebook application, and hundreds of libraries. The following are the steps you need to follow for installing the Anaconda distribution.

1. Open the following URL: `https://www.anaconda.com/products/individual`
2. Click the installer for your operating system, as shown in Figure 1-1. The installer gets downloaded to your system.

G. Rajagopalan, *A Python Data Analyst's Toolkit*, https://doi.org/10.1007/978-1-4842-6399-0_1

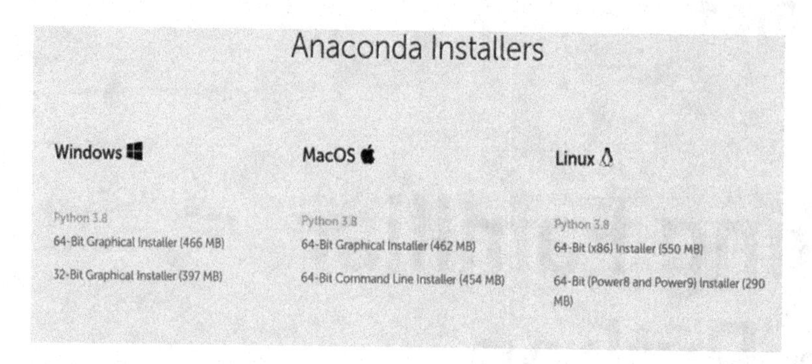

Figure 1-1. *Installing Anaconda*

3. Open the installer (file downloaded in the previous step) and run it.
4. After the installation is complete, open the Jupyter application by typing "jupyter notebook" or "jupyter" in the explorer (search bar) next to the start menu, as shown in Figure 1-2 (shown for Windows OS).

Figure 1-2. *Launching Jupyter*

Please follow the following steps for downloading all the data files used in this book:

- Click the following link: `https://github.com/DataRepo2019/Data-files`
- Select the green "Code" menu and click on "Download ZIP" from the dropdown list of this menu
- Extract the files from the downloaded zip folder and import these files into your Jupyter application

Now that we have installed and launched Jupyter, let us understand how to use this application in the next section.

Getting started with Jupyter notebooks

Before we discuss the essentials of Jupyter notebooks, let us discuss what an integrated development environment (or IDE) is. An IDE brings together the various activities involved in programming, like including writing and editing code, debugging, and creating executables. It also includes features like autocompletion (completing what the user wants to type, thus enabling the user to focus on logic and problem-solving) and syntax highlighting (highlighting the various elements and keywords of the language). There are many IDEs for Python, apart from Jupyter, including Enthought Canopy, Spyder, PyCharm, and Rodeo. There are several reasons for Jupyter becoming a ubiquitous, de facto standard in the data science community. These include ease of use and customization, support for several programming languages, platform independence, facilitation of access to remote data, and the benefit of combining output, code, and multimedia under one roof.

JupyterLab is the IDE for Jupyter notebooks. Jupyter notebooks are web applications that run locally on a user's machine. They can be used for loading, cleaning, analyzing, and modeling data. You can add code, equations, images, and markdown text in a Jupyter notebook. Jupyter notebooks serve the dual purpose of running your code as well as serving as a platform for presenting and sharing your work with others. Let us look at the various features of this application.

1. **Opening the dashboard**

   Type "jupyter notebook" in the search bar next to the start menu. This will open the Jupyter dashboard. The dashboard can be used to create new notebooks or open an existing one.

2. **Creating a new notebook**

   Create a new Jupyter notebook by selecting *New* from the upper right corner of the Jupyter dashboard and then select *Python 3* from the drop-down list that appears, as shown in Figure 1-3.

***Figure 1-3.** Creating a new Jupyter notebook*

3. **Entering and executing code**

   Click inside the first cell in your notebook and type a simple line of code, as shown in Figure 1-4. Execute the code by selecting *Run Cells* from the "Cell" menu, or use the shortcut keys *Ctrl+Enter*.

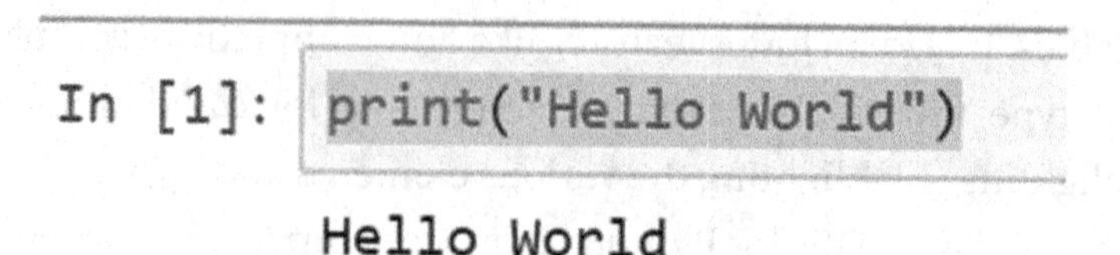

***Figure 1-4.** Simple code statement in a Jupyter cell*

4. **Adding markdown text or headings**

   In the new cell, change the formatting by selecting *Markdown* as shown in Figure 1-5, or by pressing the keys *Esc+M* on your keyboard. You can also add a heading to your Jupyter notebook by selecting *Heading* from the drop-down list shown in the following or pressing the shortcut keys *Esc+(1/2/3/4).*

***Figure 1-5.** Changing the mode to Markdown*

5. **Renaming a notebook**

   Click the default name of the notebook and type a new name, as shown in Figure 1-6.

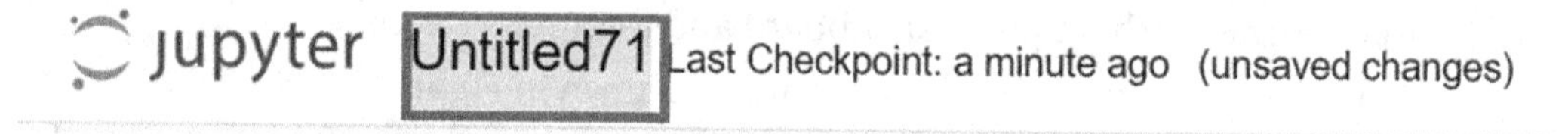

***Figure 1-6.** Changing the name of a file*

   You can also rename a notebook by selecting *File ➤ Rename.*

6. **Saving a notebook**

   Press Ctrl+S or choose File ➤ Save and Checkpoint.

7. **Downloading the notebook**

   You can email or share your notebook by downloading your notebook using the option *File ➤ Download as ➤ notebook (.ipynb),* as shown in Figure 1-7.

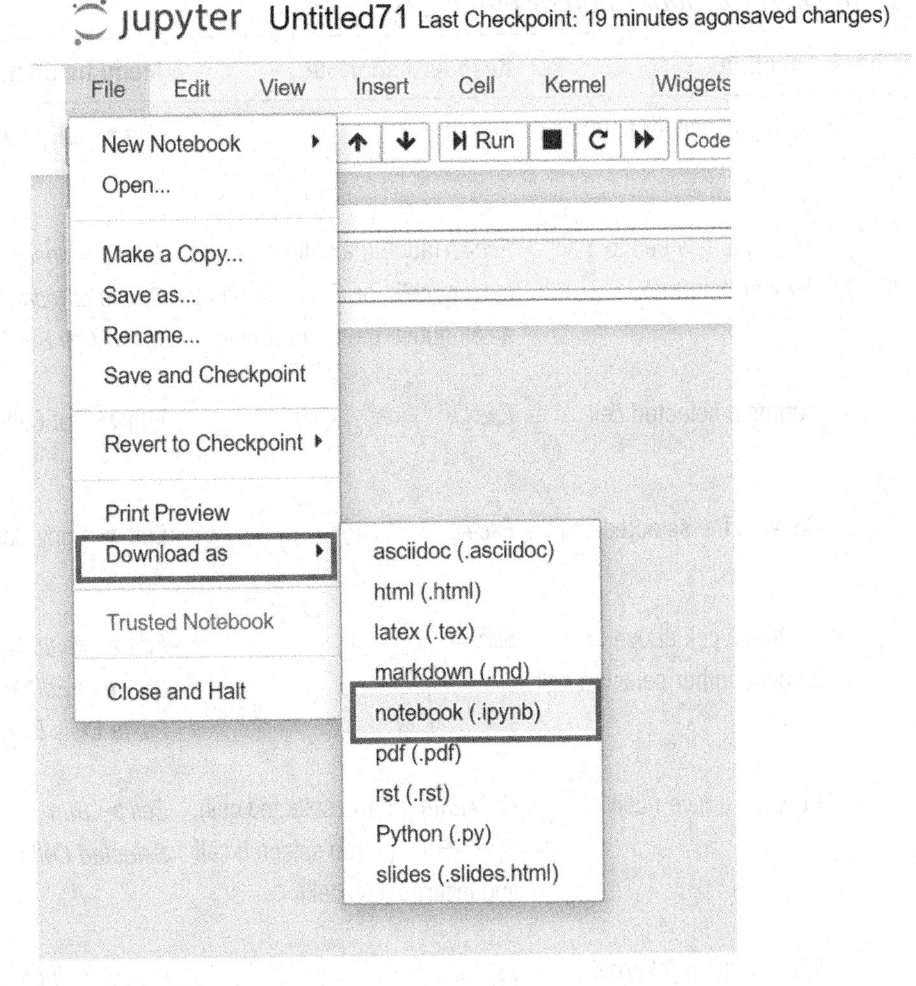

Figure 1-7. *Downloading a Jupyter notebook*

Shortcuts and other features in Jupyter

Let us look at some key features of Jupyter notebooks, including shortcuts, tab completions, and magic commands.

Table 1-1 gives some of the familiar icons found in Jupyter notebooks, the corresponding menu functions, and the keyboard shortcuts.

***Table 1-1.** Jupyter Notebook Toolbar Functions*

Icon in Toolbar	Function	Keyboard shortcut	Menu function
	Saving a Jupyter notebook	*Esc+s*	*File ➤ Save as*
	Adding a new cell to a Jupyter notebook	*Esc+b* (adding a cell below the current cell), or *Esc+a* (adding a cell above the current cell)	*Insert ➤ Insert Cell Above or Insert ➤ Insert Cell Below*
	Cutting a selected cell	*Esc+x*	Edit ➤ Cut Cells
	Copying the selected cell	*Esc+c*	Edit ➤ Copy Cells
	Pasting a cell above or below another selected cell	*Esc+v*	*Edit ➤ Paste Cells Above* or *Edit ➤ Paste Cells Below*
 Run	Running a given cell	*Ctrl+Enter* (to run selected cell); *Shift+Enter* (to run selected cell and insert a new cell)	*Cell ➤ Run Selected Cells*
	Interrupting the kernel	*Esc+ii*	*Kernel ➤ Interrupt*
	Rebooting the kernel	*Esc+00*	*Kernel ➤ Restart*

If you are not sure about which keyboard shortcut to use, go to: *Help ➤ Keyboard Shortcuts,* as shown in Figure 1-8.

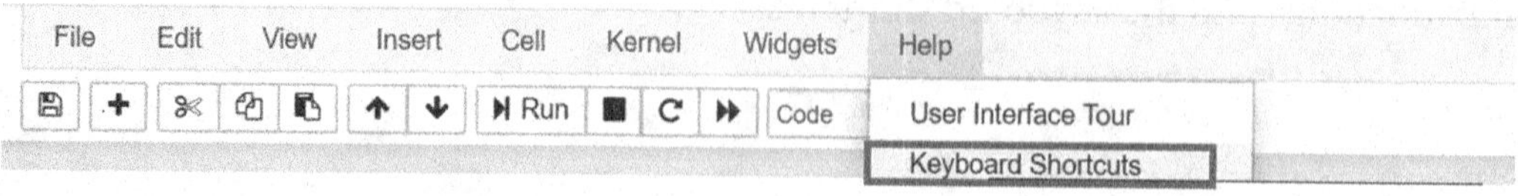

***Figure 1-8.** Help menu in Jupyter*

Commonly used keyboard shortcuts include

- *Shift+Enter* to run the code in the current cell and move to the next cell.
- *Esc* to leave a cell.
- *Esc+M* changes the mode for a cell to "Markdown" mode.
- *Esc+Y* changes the mode for a cell to "Code".

Tab Completion

This is a feature that can be used in Jupyter notebooks to help you complete the code being written. Usage of tab completions can speed up the workflow, reduce bugs, and quickly complete function names, thus reducing typos and saving you from having to remember the names of all the modules and functions.

For example, if you want to import the Matplotlib library but don't remember the spelling, you could type the first three letters, mat, and press Tab. You would see a drop-down list, as shown in Figure 1-9. The correct name of the library is the second name in the drop-down list.

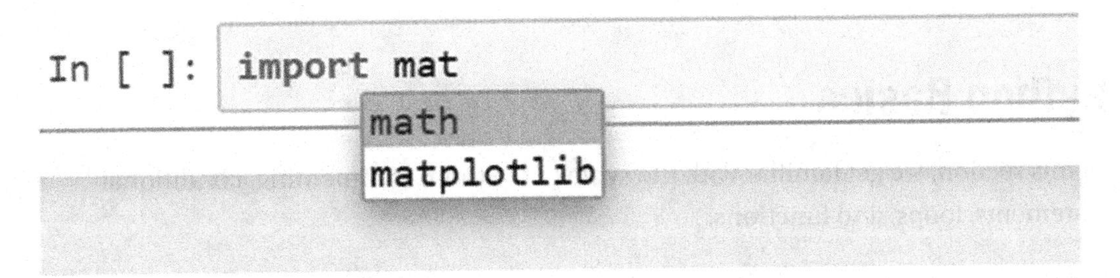

Figure 1-9. *Tab completion in Jupyter*

Magic commands used in Jupyter

Magic commands are special commands that start with one or more % signs, followed by a command. The commands that start with one % symbol are applicable for a single line of code, and those beginning with two % signs are applicable for the entire cell (all lines of code within a cell).

One commonly used magic command, shown in the following, is used to display Matplotlib graphs inside the notebook. Adding this magic command avoids the need to call the *plt.show* function separately for showing graphs (the Matplotlib library is discussed in detail in Chapter 7).

CODE:

```
%matplotlib inline
```

Magic commands, like *timeit,* can also be used to time the execution of a script, as shown in the following.

CODE:

```
%%timeit
for i in range(100000):
    i*i
```

Output:

```
16.1 ms ± 283 µs per loop (mean ± std. dev. of 7 runs, 100 loops each)
```

Now that you understand the basics of using Jupyter notebooks, let us get started with Python and understand the core aspects of this language.

Python Basics

In this section, we get familiar with the syntax of Python, commenting, conditional statements, loops, and functions.

Comments, print, and input

In this section, we cover some basics like printing, obtaining input from the user, and adding comments to help others understand your code.

Comments

A comment explains what a line of code does, and is used by programmers to help others understand the code they have written. In Python, a comment starts with the # symbol.

Proper spacing and indentation are critical in Python. While other languages like Java and C++ use brackets to enclose blocks of code, Python uses an indent of four spaces to specify code blocks. One needs to take care of indents to avoid errors. Applications like Jupyter generally take care of indentation and automatically add four spaces at the beginning of a block of code.

Printing

The *print* function prints content to the screen or any other output device.

Generally, we pass a combination of strings and variables as arguments to the print function. Arguments are the values included within the parenthesis of a function, which the function uses for producing the result. In the following statement, “Hello!” is the argument to the *print* function.

CODE:

```
print("Hello!")
```

To print multiple lines of code, we use triple quotes at the beginning and end of the string, for example:

CODE:

```
print('''Today is a lovely day.
It will be warm and sunny.
It is ideal for hiking.''')
```

Output:

```
Today is a lovely day.
It will be warm and sunny.
It is ideal for hiking.
```

Note that we do not use semicolons in Python to end statements, unlike some other languages.

The *format* method can be used in conjunction with the *print* method for embedding variables within a string. It uses curly braces as placeholders for variables that are passed as arguments to the method.

Let us look at a simple example where we print variables using the *format* method.

CODE:

```
weight=4.5
name="Simi"
print("The weight of {} is {}".format(name,weight))
```

Output:

```
The weight of Simi is 4.5
```

The preceding statement can also be rewritten as follows without the format method:

CODE:

```
print("The weight of",name,"is","weight")
```

Note that only the string portion of the print argument is enclosed within quotes. The name of the variable does not come within quotes. Similarly, if you have any constants in your print arguments, they also do not come within quotes. In the following example, a Boolean constant (True), an integer constant (1), and strings are combined in a print statement.

CODE:

```
print("The integer equivalent of",True,"is",1)
```

Output:

```
The integer equivalent of True is 1
```

The format fields can specify precision for floating-point numbers. Floating-point numbers are numbers with decimal points, and the number of digits after the decimal point can be specified using format fields as follows.

CODE:

```
x=91.234566
print("The value of x upto 3 decimal points is {:.3f}".format(x))
```

Output:

```
The value of x upto 3 decimal points is 91.235
```

We can specify the position of the variables passed to the method. In this example, we use position "1" to refer to the second object in the argument list, and position "0" to specify the first object in the argument list.

CODE:

```
y='Jack'
x='Jill'
print("{1} and {0} went up the hill to fetch a pail of water".format(x,y))
```

Output:

```
Jack and Jill went up the hill to fetch a pail of water
```

Input

The *input* function accepts inputs from the user. The input provided by the user is stored as a variable of type *String*. If you want to do any mathematical calculations with any numeric input, you need to change the data type of the input to int or float, as follows.

CODE:

```
age=input("Enter your age:")
print("In 2010, you were",int(age)-10,"years old")
```

Output:

```
Enter your age:76
In 2010, you were 66 years old
```

Further reading on Input/Output in Python: `https://docs.python.org/3/tutorial/inputoutput.html`

Variables and Constants

A constant or a literal is a value that does not change, while a variable contains a value can be changed. We do not have to declare a variable in Python, that is, specify its data type, unlike other languages like Java and C/C++. We define it by giving the variable a name and assigning it a value. Based on the value, a data type is automatically assigned to it. Values are stored in variables using the assignment operator (=). The rules for naming a variable in Python are as follows:

- a variable name cannot have spaces
- a variable cannot start with a number

- a variable name can contain only letters, numbers, and underscore signs (_)
- a variable cannot take the name of a reserved keyword (for example, words like *class, continue, break, print*, etc., which are predefined terms in the Python language, have special meanings, and are invalid as variable names)

Operators

The following are some commonly used operators in Python.

Arithmetic operators: Take two integer or float values, perform an operation, and return a value.

The following arithmetic operators are supported in Python:

- **(Exponent)
- %(modulo or remainder),
- //(quotient),
- *(multiplication)
- -(subtraction)
- +(addition)

The order of operations is essential. Parenthesis takes precedence over exponents, which takes precedence over division and multiplication, which takes precedence over addition and subtraction. An acronym was designed - P.E.D.M.A.S.(Please Excuse My Dear Aunt Sally) - that can be used to remember the order of these operations to understand which operator first needs to be applied in an arithmetic expression. An example is given in the following:

CODE:

```
(1+9)/2-3
```

Output:

```
2.0
```

In the preceding expression, the operation inside the parenthesis is performed first, which gives 10, followed by division, which gives 5, and then subtraction, which gives the final output as 2.

Comparison operators: These operators compare two values and evaluate to a true or false value. The following comparison operators are supported in Python:

- >: Greater than
- < : Less than
- <=: Less than or equal to
- >=: Greater than or equal to
- == : equality. Please note that this is different from the assignment operator (=)
- !=(not equal to)

Logical (or Boolean) operators: Are similar to comparison operators in that they also evaluate to a *true* or *false* value. These operators operate on Boolean variables or expressions. The following logical operators are supported in Python:

- *and operator*: An expression in which this operator is used evaluates to *True* only if all its subexpressions are *True*. Otherwise, if any of them is *False*, the expression evaluates to *False*
  An example of the usage of the *and* operator is shown in the following.
  CODE:

  ```
  (2>1) and (1>3)
  ```

  Output:

  ```
  False
  ```

- *or* operator: An expression in which the *or* operator is used, evaluates to *True* if any one of the subexpressions within the expression is *True*. The expression evaluates to *False* if all its subexpressions evaluate to *False*.

  An example of the usage of the *or* operator is shown in the following.

  CODE:

  ```
  (2>1) or (1>3)
  ```

Output:

```
True
```

- *not* operator: An expression in which the *not* operator is used, evaluates to *True* if the expression is *False*, and vice versa.

  An example of the usage of the *not* operator is shown in the following.
  CODE:

  ```
  not(1>2)
  ```

  Output:

  ```
  True
  ```

Assignment operators

These operators assign a value to a variable or an operand. The following is the list of assignment operators used in Python:

- = (assigns a value to a variable)
- += (adds the value on the right to the operand on the left)
- -= (subtracts the value on the right from the operand on the left)
- *= (multiplies the operand on the left by the value on the right)
- %= (returns the remainder after dividing the operand on the left by the value on the right)
- /= (returns the quotient, after dividing the operand on the left by the value on the right)
- //= (returns only the integer part of the quotient after dividing the operand on the left by the value on the right)

Some examples of the usage of these assignment operators are given in the following.

CODE:

```
x=5 #assigns the value 5 to the variable x
x+=1 #statement adds 1 to x (is equivalent to x=x+1)
x-=1 #statement subtracts 1 from x (is equivalent to x=x-1)
x*=2 #multiplies x by 2(is equivalent to x=x*2)
```

```
x%=3 #equivalent to x=x%3, returns remainder
x/=3 #equivalent to x=x/3, returns both integer and decimal part of quotient
x//=3 #equivalent to x=x//3, returns only the integer part of quotient
after dividing x by 3
```

Identity operators (is and not is)

These operators check for the equality of two objects, that is, whether the two objects point to the same value and return a Boolean value (*True/False*) depending on whether they are equal or not. In the following example, the three variables "*x*", "*y*", and "*z*" contain the same value, and hence, the identity operator (*is*) returns *True* when "x" and "z" are compared.

Example:

```
x=3
y=x
z=y
x is z
```

Output:

```
True
```

Membership operators (in and not in)

These operators check if a particular value is present in a string or a container (like lists and tuples, discussed in the next chapter). The *in* operator returns "True" if the value is present, and the *not in* operator returns "True" if the value is not present in the string or container.

CODE:

```
'a' in 'and'
```

Output:

```
True
```

Data types

The data type is the category or the type of a variable, based on the value it stores.

The data type of a variable or constant can be obtained using the *type* function.

CODE:

```
type(45.33)
```

Output:

```
float
```

Some commonly used data types are given in Table 1-2.

Table 1-2. *Common Data Types in Python*

Type of data	Data type	Examples

Numeric data	*int*: for numbers without a decimal point
	float: for numbers with a decimal point
		`#int`
		`a=1`
		`#float`
		`b=2.4`

Sequences	Sequences store more than one value. Some of the sequences in Python are:
	• *list*
	• *range*
	• *tuple*
		`#tuple`
		`a=(1,2,3)`
		`#list`
		`b=[1,2,3]`
		`#range`
		`c=range(5)`

Characters or text	*str* is the data type for storing a single character or a sequence of characters within quotes	`#single character`
		`X='a'`
		`#multiple characters`
		`x='hello world'`
		`#multiple lines`
		`x='''hello world`
		`good morning'''`

Boolean data	*bool* is the data type for storing True or False values	`X=True` `Y=False`
Mapping objects	*dict* is the data type for a dictionary (an object mapping a key to a value)	`x={'Apple':'fruit',` `'Carrot':'vegetable'}`

Representing dates and times

Python has a module called *datetime* that allows us to define a date, time, or duration.

We first need to import this module so that we can use the functions available in this module for defining a date or time object, using the following statement.

CODE:

```
import datetime
```

Let us use the methods that are part of this module to define various date/time objects.

Date object

A date consisting of a day, month, and year can be defined using the *date* method, as shown in the following.

CODE:

```
date=datetime.date(year=1995,month=1,day=1)
print(date)
```

Output:

```
1995-01-01
```

Note that all three arguments of the *date* method - day, month, and year - are mandatory. If you skip any of these arguments while defining a *date* object, an error occurs, as shown in the following.

CODE:

```
date=datetime.date(month=1,day=1)
print(date)
```

Output:

```
TypeError                                 Traceback (most recent call last)
<ipython-input-3-7da76b18c6db> in <module>
----> 1 date=datetime.date(month=1,day=1)
      2 print(date)

TypeError: function missing required argument 'year' (pos 1)
```

Time object

To define an object in Python that stores time, we use the *time* method.

The arguments that can be passed to this method may include hours, minutes, seconds, or microseconds. Note that unlike the *date* method, arguments are not mandatory for the *time* method (they can be skipped).

CODE:

```
time=datetime.time(hour=12,minute=0,second=0,microsecond=0)
print("midnight:",time)
```

Output:

```
midnight: 00:00:00
```

Datetime object

We can also define a datetime object consisting of both a date and a time, using the *datetime* method, as follows. For this method, the date arguments – day, month, and year – are mandatory, but the time argument (like hour, minute, etc.) can be skipped.

CODE:

```
datetime1=datetime.datetime(year=1995,month=1,day=1,hour=12,minute=0,second
=0,microsecond=0)
print("1st January 1995 midnight:", datetime1)
```

Output:

```
1st January 1995 midnight: 1995-01-01 12:00:00
```

Timedelta object

A *timedelta* object represents a specific duration of time, and is created using the *timedelta* method.

Let us create a *timedelta* object that stores a period of 17 days.

CODE:

```
timedelta1=datetime.timedelta(weeks=2,days=3)
timedelta1
```

Output:

```
datetime.timedelta(days=17)
```

You can also add other arguments like seconds, minutes, and hours, while creating a *timedelta* object.

A *timedelta* object can be added to an existing date or datetime object, but not to a time object

Adding a duration (*timedelta* object) to a *date* object:

CODE:

```
#adding a duration to a date object is supported
date1=datetime.date(year=1995,month=1,day=1)
timedelta1=datetime.timedelta(weeks=2,days=3)
date1+timedelta1
```

Output:

```
datetime.date(1995, 1, 18)
```

Adding a duration (*timedelta* object) to a *datetime* object:

CODE:

```
#adding a duration to a datetime object is supported
datetime1=datetime.datetime(year=1995,month=2,day=3)
timedelta1=datetime.timedelta(weeks=2,days=3)
datetime1+timedelta1
```

Output:

```
datetime.datetime(1995, 2, 20, 0, 0)
```

Adding a duration to a *time* object leads to an error:

CODE:

```
#adding a duration to a time object is not supported
time1=datetime.time(hour=12,minute=0,second=0,microsecond=0)
timedelta1=datetime.timedelta(weeks=2,days=3)
time1+timedelta1
```

Output:

```
TypeError                                 Traceback (most recent call last)
<ipython-input-9-5aa64059a69a> in <module>
      2 time1=datetime.time(hour=12,minute=0,second=0,microsecond=0)
      3 timedelta1=datetime.timedelta(weeks=2,days=3)
----> 4 time1+timedelta1

TypeError: unsupported operand type(s) for +: 'datetime.time' and
'datetime.timedelta'
```

Further reading:
Learn more about the Python datetime module
https://docs.python.org/3/library/datetime.html

Working with Strings

A string is a sequence of one or more characters enclosed within quotes (both single and double quotes are acceptable). The data type for strings is *str*. Python does not support the character data type, unlike older languages like Java and C. Even single characters, like 'a', 'b', are stored as strings. Strings are internally stored as arrays and are immutable (cannot be modified). Let us see how to define a string.

Defining a string

Single-line strings can be defined using single or double quotes.

CODE:

```
x='All that glitters is not gold'
#OR
x="All that glitters is not gold"
```

For multiline strings, use triple quotes:

CODE:

```
x='''Today is Tuesday.
Tomorrow is Wednesday'''
```

String operations

Various functions can be used with strings, some of which are explained in the following.

1. Finding the length of a string: The *len* function can be used to calculate the length of a string, as shown in the following.

   CODE:

   ```
   len('Hello')
   ```

   Output:

   ```
   5
   ```

2. Accessing individual elements in a string:

   The individual characters in a string can be extracted using the indexing operator, [].

   CODE:

   ```
   x='Python'
   x[3]
   Output:
   'h'
   ```

3. Slicing a string: Slicing refers to the extraction of a portion or subset of an object (in this case, the object is a string). Slicing can also be used with other iterable objects like lists and tuples, which we discuss in the next chapter. The colon operator is used for slicing, with an optional start, stop, and step index. Some examples of slicing are provided in the following.

   CODE:

   ```
   x='Python'

   x[1:] #from second character to the end
   ```

   Output:

   ```
   'ython'
   ```

Some more examples of slicing:

CODE:

```
x[:2] #first two characters. The starting index is assumed to be 0
```

Output:

```
'Py'
```

CODE:

```
x[::-1]#reversing the string, the last character has an index -1
```

Output:

```
'nohtyP'
```

4. Justification:

   To add spaces to the right or left, or center the string, the *rjust, ljust,* or *center* method is used. The first argument passed to such a method is the length of the new string, and the optional second argument is the character to be used for padding. By default, spaces are used for padding.

   CODE:

   ```
   '123'.rjust(5,"*")
   ```

   Output:

   ```
   '**123'
   ```

5. Changing the case: To change the case of the string, the *upper* or *lower* method is used, as shown in the following.
   CODE:

   ```
   'COLOR'.lower()
   ```

   Output:

   ```
   'color'
   ```

6. Checking what a string contains:

   In order to check whether a string starts or ends with a given character, the *startswith* or *endswith* method is used.
   CODE:

   ```
   'weather'.startswith('w')
   ```

   Output:

   ```
   True
   ```

7. Removing whitespaces from a string:

   To remove spaces from a string, use the *strip* method (to remove spaces at both ends), *rstrip* (to remove spaces from the right end), or the *lstrip* method (to remove spaces from the left end). An example is shown in the following.

   CODE:

   ```
   '  Hello'.lstrip()
   ```

   Output:

   ```
   'Hello'
   ```

8. Examining the contents of a string:

   There are several methods to check what a string contains, like `isalpha, isupper, isdigit, isalnum`, etc. All these methods return "True" only if all the characters in the string satisfy a given condition.

   CODE:

   ```
   '981'.isdigit()#to check for digits
   ```

   Output:

   ```
   True
   ```

CODE:

```
'Abc'.isupper()
#Checks if all characters are in uppercase. Since all letters are
not uppercase, the condition is not satisfied
```

Output:

```
False
```

9. Joining a list of strings:

   The *join* method combines a list of strings into one string. On the left-hand side of the *join* method, we mention the delimiter in quotes to be used for joining the strings. On the right-hand side, we pass the list of individual strings.

   CODE:

   ```
   ' '.join(['Python','is','easy','to','learn'])
   ```

   Output:

   ```
   'Python is easy to learn'
   ```

10. Splitting a string:

    The *split* method does the opposite of what the join method does. It breaks down a string into a list of individual words and returns the list. If we just pass one word to this method, it returns a list containing just one word and does not split the string further.

    CODE:

    ```
    'Python is easy to learn'.split()
    ```

    Output:

    ```
    ['Python', 'is', 'easy', 'to', 'learn']
    ```

Conditional statements

Conditional statements, as the name indicates, evaluate a condition or a group of conditions. In Python, the *if-elif-else* construct is used for this purpose. Python does not have the switch-case construct, which is used in some other languages for conditional execution.

Conditional statements start with the *if* keyword, and the expression or a condition to be evaluated. This is followed by a block of code that executes only if the condition evaluates to "True"; otherwise it is skipped.

The *else* statement (which does not contain any condition) is used to execute a block of code when the condition mentioned in the *if* statement is not satisfied. The *elif* statements are used to evaluate specific conditions. The order of *elif* statements matters. If one of the *elif* statements evaluates to True, the *elif* statements following it are not executed at all. The *if* statement can also exist on its own, without mentioning the *else* or *elif* statements.

The following example demonstrates the *if-elif-else* construct.

CODE:

```
#if-elif-else
color=input('Enter one of the following colors - red, orange or blue:')
if color=='red':
    print('Your favorite color is red')
elif color=='orange':
    print('Your favorite color is orange')
elif color=='blue':
    print('Your favorite color is blue')
else:
    print("You entered the wrong color")
```

Output:

```
Enter one of the following colors - red, orange or blue:pink
You entered the wrong color
```

Conditional statements can be nested, which means that we can have one conditional statement (inner) within another (outer). You need to be particularly careful with indentation while using nested statements. An example of nested *if* statements is shown in the following.

CODE:

```
#nested conditionals
x=20
if x<10:
    if x<5:
        print("Number less than 5")
    else:
        print("Number greater than 5")
else:
    print("Number greater than 10")
```

Output:

```
Number greater than 10
```

Further reading: See more on the *if* statement: https://docs.python.org/3/tutorial/controlflow.html#if-statements

Loops

Loops are used to execute a portion of the code repeatedly. A single execution of a block of code is called an iteration, and loops often go through multiple rounds of iterations. There are two types of loops that are used in Python – the *for* loop and the *while* loop.

While loop

The *while* loop is used when we want to execute particular instructions as long as a condition is "True". After the block of code executes, the execution goes back to the beginning of the block. An example is shown in the following.

CODE:

```
#while loop with continue statement
while True:
    x=input('Enter the correct color:')
    if(x!='red'):
        print("Color needs to be entered as red")
        continue
    else:
        break
```

Output:

```
Enter the correct color:blue
Color needs to be entered as red
Enter the correct color:yellow
Color needs to be entered as red
Enter the correct color:red
```

In the preceding example, the first statement (*while True*) is used to execute an infinite loop. Once the username entered is of the right length, the break statement takes execution outside the loop; otherwise, a message is displayed to the user asking for a username of the right length. Note that execution automatically goes to the beginning of the loop, after the last statement in the block of code.

The *break* statement is used to take the control outside the loop. It is useful when we have an infinite loop that we want to break out of.

The *continue* statement does the opposite - it takes control to the beginning of the loop. The keywords *break* and *continue* can be used both with loops and conditional statements, like *if/else*.

Further reading:

See more about the following:

- *break* and *continue* statements: `https://docs.python.org/3/tutorial/controlflow.html#break-and-continue-statements-and-else-clauses-on-loops`
- *while* statement: `https://docs.python.org/3/reference/compound_stmts.html#while`

for loop

The *for* loop is used to execute a block of a code a predetermined number of times. The *for* loop can be used with any kind of iterable object, that is, a sequence of values that can be used by a loop for running repeated instances or iterations. These iterable objects include lists, tuples, dictionaries, and strings.

The *for* loop is also used commonly in conjunction with the *range* function. The range function creates a *range* object, another iterable, which is a sequence of evenly spaced integers. Consider the following example where we calculate the sum of the first five odd integers using a *for* loop.

CODE:

```
#for loop
sum=0
for i in range(1,10,2):
    sum=sum+i
print(sum)
```

Output:

```
25
```

The *range* function has three arguments: the start argument, the stop argument, and the step argument. None of these three arguments are mandatory. Numbers from 0 to 9 (both 0 and 9 included) can be generated as *range(10)*, *range(0,10)*, or *range(0,10,1)*. The default start argument is 0, and the default step argument is 1.

For loops can also be nested (with an outer loop and any number of inner loops), as shown in the following.

CODE:

```
#nested for loop
for i in 'abcd':
    for j in range(4):
        print(i,end=" ")
    print("\n")
```

Output:

```
a a a a

b b b b

c c c c

d d d d
```

Further reading: See more about the for statement: `https://docs.python.org/3/tutorial/controlflow.html#for-statements`

Functions

A function can be thought of as a "black box" (the user need not be concerned with the internal workings of the function) that takes an input, processes it, and produces an output. A function is essentially a block of statements performing a specific task.

In Python, a function is defined using the *def* keyword. This is followed by the name of a function and one or more optional parameters. A parameter is a variable that exists only within a function. Variables defined within a function have local scope, which means that they cannot be accessed outside the function. They are also called local variables. External code or functions cannot manipulate the variables defined within a function.

A function may have an optional return value. The return value is the output produced by a function that is returned to the main program. Calling a function means giving the function inputs (arguments) to perform its task and produce an output.

The utility of functions lies in their reusability. They also help in avoiding redundancy and organizing code into logical blocks. We just need to supply it with the set of inputs it needs to run the instructions. A function can be called repeatedly instead of manually typing out the same lines of code.

For example, say you want to find out the prime numbers in a given list of numbers. Once you have written a function for checking whether an integer is a prime number, you can simply pass each number in the list as an argument to the function and call it, instead of writing the same lines of code for each integer you want to test.

CODE:

```
def checkPrime(i):
     #Assume the number is prime initially
    isPrime=True
    for j in range(2,i):
        # checking if the number is divisible by any number between 2 and i
        if i%j==0:
            #If it is divisible by any number in the j range, it is not prime
            isPrime=False
    # This is the same as writing if isPrime==True
    if isPrime:
        print(i ,"is prime")
```

```
    else:
        print(i, "is not prime")
for i in range(10,20):
    checkPrime(i)
```

Output:

```
10 is not prime
11 is prime
12 is not prime
13 is prime
14 is not prime
15 is not prime
16 is not prime
17 is prime
18 is not prime
19 is prime
```

Further reading: See more about defining functions: `https://docs.python.org/3/tutorial/controlflow.html#defining-functions`

Anonymous or lambda functions are defined using the *lambda* keyword. They are single-expression functions and provide a compact way of defining a function without binding the function object to a name. The reason these functions are called "anonymous" is that they do not need a name. Consider the following example where we use a lambda function to calculate the sum of two numbers.

CODE:

```
(lambda x,y:(x+y))(5,4)
```

Output:

```
9
```

Note the syntax of an anonymous function. It starts with the *lambda* keyword, followed by the parameters ('x' and 'y', in this case). Then comes the colon, after which there is an expression that is evaluated and returned. There is no need to mention a return statement since there is an implicit return in such a function. Notice that the function also does not have a name.

Syntax errors and exceptions

Syntax errors are errors that may be committed inadvertently by the user while writing the code, for example, spelling a keyword wrong, not indenting the code, and so on. An exception, on the other hand, is an error that occurs during program execution. A user may enter incorrect data while running the program. If you want to divide a number (say, 'a') by another number (say, 'b'), but give a value of 0 to the denominator ('b'), this will generate an exception. The exceptions, which are autogenerated in Python and displayed to the user, may not lucidly convey the problem. Using exception handling with the *try-except* construct, we can frame a user-friendly message to enable the user to better correct the error.

There are two parts to exception handling. First, we put the code that is likely to cause an error under a *try* clause. Then, in the *except* clause, we try to deal with whatever caused an error in the *try* block. We mention the name of the exception class in the *except* clause, followed by a code block where we handle the error. A straightforward method for handling the error is printing a message that gives the user more details on what they need to correct.

Note that all exceptions are objects that are derived from the class *BaseException*, and follow a hierarchy.

Further reading: The class hierarchy for exceptions in Python can be found here: `https://docs.python.org/3/library/exceptions.html#exception-hierarchy`

A simple example of a program, with and without exception handling, is shown below.

```
while True:
    try:
        n=int(input('Enter your score:'))
        print('You obtained a score of ',n)
        break
    except ValueError:
        print('Enter only an integer value')
```

Output:

```
Enter your score(without a decimal point):abc
Enter only an integer value
Enter your score(without a decimal point):45.5
```

```
Enter only an integer value
Enter your score(without a decimal point):90
You obtained a score of  90
```

Same program (Without exception handling):

CODE:

```
n=int(input('Enter your score:'))
print('You obtained a score of ',n)
```

Output:

```
Enter your score:ninety three
---------------------------------------------------------------------------
ValueError                                Traceback (most recent call last)
<ipython-input-12-aa4fbda9d45f> in <module>
----> 1 n=int(input('Enter your score:'))
      2 print('You obtained a score of ',n)

ValueError: invalid literal for int() with base 10: 'ninety three'
```

The statement that is likely to cause an error in the preceding code is: `int(input('Enter your score:'))`. The *int* function requires an integer as an argument. If the user enters a floating-point or string value, a *ValueError* exception is generated. When we use the *try-except* construct, the *except* clause prints a message asking the user to correct the input, making it much more explicit.

Working with files

We can use methods or functions in Python to read or write to files. In other words, we can create a file, add content or text to it, and read its contents by using the methods provided by Python.

Here, we discuss how to read and write to comma-separated value (CSV) files. CSV files or comma-separated files are text files that are a text version of an Excel spreadsheet.

The functions for all of these operations are defined under the CSV module. This module has to be imported first, using the `import csv` statement, to use its various methods.

Reading from a file

Reading from a file from Python involves the following steps:

1. Using the `with open` statement, we can open an existing CSV file and assign the resulting file object to a variable or file handle (named 'f' in the following example). Note that we need to specify the path of the file using either the absolute or relative path. After this, we need to specify the mode for opening the file. For reading, the mode is 'r'. The file is opened for reading by default if we do not specify a mode.
2. Following this, there is a block of code that starts with storing the contents of the file in a read object, using the *csv.reader* function where we specify the file handle, f, as an argument.
3. However, the contents of this file are not directly accessible through this read object. We create an empty list (named 'contents' in the following example), and then we loop through the read object we created in step 2 line by line using a for loop and append it to this list. This list can then be printed to view the lines of the CSV file we created.

CODE:

```
#Reading from a file
import csv
with open('animals.csv') as f:
    contents=csv.reader(f)
    lines_of_file=[]
    for line in contents:
        lines_of_file+=line
lines_of_file
```

Writing to a file

Writing to a file involves the following steps.

1. Using the *open* function, open an existing CSV file or if the file does not exist, the open function creates a new file. Pass the name of the file (with the absolute path) in quotes and specify the mode as 'w', if you want to overwrite the contents or write into a new file. Use the 'a' or 'append' mode if you simply want to append some lines to an existing file. Since we do not want to overwrite in this case, we open the file using the append ('a') mode. Store it in a variable or file handle and give it a name, let us say 'f'.

2. Using the *csv.writer()* function, create a writer object to add the content since we cannot directly write to the CSV file. Pass the variable (file handle), 'f', as an argument to this function.

3. Invoke the *writerow* method on the writer object created in the previous step. The argument to be passed to this method is the new line to be added (as a list).

4. Open the CSV file on your system to see if the changes have been reflected.

CODE:

```
#Writing to a file
with open(r'animals.csv',"w") as f:
    writer_object=csv.writer(f,delimiter=",")
    writer_object.writerow(['sheep','lamb'])
```

The modes that can be used with the open function to open a file are:

- "r": opens a file for only reading.
- "w": opens a file for only writing. It overwrites the file if it already exists.
- "a": opens a file for writing at the end of the file. It retains the original file contents.
- "w+": opens the file for both reading and writing.

Further reading: See more about reading and writing to files in Python: `https://docs.python.org/3/tutorial/inputoutput.html#reading-and-writing-files`

Modules in Python

A module is a Python file with a .py extension. It can be thought of as a section of a physical library. Just as each section of a library (for instance, fiction, sports, fitness) contains books of a similar nature, a module contains functions that are related to one another. For example, the *matplotlib* module contains all functions related to plotting graphs. A module can also contain another module. The *matplotlib* module, for instance, contains a module called *pyplot*. There are many built-in functions in Python that are part of the standard library and do not require any module to be imported to use them.

A module can be imported using the *import* keyword, followed by the name of the module:

CODE:

```
import matplotlib
```

You can also import part of a module (a submodule or a function) using the *from* keyword. Here, we are importing the cosine function from the math module:

CODE:

```
from math import cos
```

Creating and importing a customized module in Python requires the following steps:

1. Type "idle" in the search bar next to the start menu. Once the Python shell is open, create a new file by selecting: *File* ➤ *New File*

2. Create some functions with similar functionality. As an example, here, we are creating a simple module that creates two functions - sin_angle and cos_angle. These functions calculate the sin and cosine of an angle (given in degrees).

   CODE:

   ```
   import math
   def sin_angle(x):
       y=math.radians(x)
       return math.sin(y)
   ```

```
def cos_angle(x):
    y=math.radians(x)
    return math.cos(y)
```

3. Save the file. This directory, where the file should be saved, is the same directory where Python runs. You can obtain the current working directory using the following code:

   CODE:

   ```
   import os
   os.getcwd()
   ```

4. Using the import statement, import and use the module you have just created.

Python Enhancement Proposal (PEP) 8 – standards for writing code

Python Enhancement Proposal (PEP) is a technical document that provides documentation for new features introduced in the Python language. There are many types of PEP documents, the most important among these being PEP 8. The PEP 8 document provides style guidelines for writing code in Python. The main emphasis of PEP 8 is on providing a set of consistent rules that enhance code readability – anybody who reads your code should be able to understand what you are doing. You can find the complete PEP8 document here: `https://www.python.org/dev/peps/pep-0008/`

There are several guidelines in PEP8 for different aspects of the code, some of which are summarized in the following.

- Indentation: Indentation is used to indicate the starting of a block of code. In Python, four spaces are used for indentation. Tabs should be avoided for indentation.
- Line length: The maximum character length for a line of code is 79 characters. For comments, the limit is 72 characters.

- The naming conventions for naming different types of objects in Python are also laid out in PEP 8. Short names should be used, and underscores can be used to improve readability. For naming functions, methods, variables, modules, and packages, use the lowercase (all small letters) notation. With constants, use the uppercase (all capitals) notation, and for classes, use the CapWords (two words each starting with a capital letter, not separated by spaces) notation for naming.
- Comments: Block comments, starting with a # and describing an entire block of code, are recommended. Inline comments, which are on the same line as the line of code, as shown in the following, should be avoided. If they are used at all, they should be separated by two spaces from the code.

  CODE:

  ```
  sum+=1 #incrementing the sum by 1
  ```

- Imports:

  While importing a module to use it in your code, avoid wildcard imports (using the * notation), like the one shown in the following.

  CODE:

  ```
  from module import *
  ```

  Multiple packages or classes should not be imported on the same line.

  CODE:

  ```
  import a,b
  ```

  They should be imported on separate lines, as shown in the following.

  CODE:

  ```
  import a
  import b
  ```

Absolute imports should be used as far as possible, for example:

CODE:

```
import x.y
```

Alternatively, we can use this notation for importing modules:

CODE:

```
from x import y
```

- Encoding: The encoding format to be used for writing code in Python 3 is UTF-8

Summary

- The syntax of Python differs from other languages like Java and C, in that statements in Python do not end with a semicolon, and spaces (four spaces), are used for indentation, instead of curly braces.
- Python has basic data types like *int, float, str,* and *bool,* among many others, and operators (arithmetic, Boolean, assignment, and comparison) that operate on variables depending on their data type.
- Python has the *if-elif-else* keywords for the conditional execution of statements. It also has the *for* loop and the *while* loop for repeating a specific portion of the program.
- Functions help with reusing a part of code and avoiding redundancy. Each function should perform only one task. Functions in Python are defined using the *def* keyword. Anonymous or *lambda* functions provide a shortcut for writing functions in a single line without binding the function to a name.
- A module is a collection of similar functions and is a simple Python file. Apart from the functions that are part of the standard library, any function that is part of an external module requires the module to be imported using the *import* keyword.

- Python has functions for creating, reading, and writing to text and CSV files. The files can be opened in various modes, depending on whether you want to read, write, or append data.
- Exception handling can be used to handle exceptions that occur during the execution of the program. Using the *try* and *except* keywords, we can deal with the part of the program that is likely to cause an exception.
- PEP 8 sets standards for a range of coding-related aspects in Python, including usage of spaces, tabs, and blank lines, and conventions for naming and writing comments.

The next chapter delves deep into topics like containers, like lists, tuples, dictionaries, and sets. We also discuss a programming paradigm known as object-oriented programming, and how it is implemented using classes and objects.

Review Exercises

Question 1

Calculate the factorial of numbers from 1 to 5 using nested loops.

Question 2

A function is defined using which of the following?

1. *def* keyword
2. *function* keyword
3. *void* keyword
4. No keyword is required

Question 3

What is the output of the following code?

```
x=True
y=False
z=True
x+y+z
```

Question 4

Write a Python program to print the following sequence:

*****

****

***

**

*

Question 5

Which of these variables has been defined using the correct syntax?

1. `1x=3`
2. `x 3=5`
3. `x`
4. `x_3=5`
5. `x$4=5`

Question 6

What is the output of the following code? (Hint: The id function returns the memory address of an object.)

```
str1="Hello"
str2=str1
id(str1)==id(str2)
```

Question 7

Convert the string "123-456-7890" into the format "1234567890". Use the *join* and *split* string functions.

Question 8

Write a function that performs the following tasks (the name of the file is passed as a parameter):

- Create a new text file with the name passed as an argument to the function
- Add a line of text ("Hello World") to the file
- Read the contents of the file
- Opens the file again, add another line ("This is the next line") below the first line
- Reread the file and print the contents on the screen

Answers

Question 1

Solution:

```
#Question 1
for i in range(1,6):
    fact=1
    for j in range(1,i+1):
        fact=fact*j
    print("Factorial of number ",i," is:",fact)
```

Question 2

Option 1: Functions in Python require the *def* keyword.

Question 3

Output: 2

Explanation: The Boolean value "True" is treated as value 1, and 'False' as value 0. Applying the addition operator on Boolean variables is valid.

Question 4

Solution

```
#question 4
l=range(6)
for i in l[::-1]:
    print("*"*i)
    print("\n")
```

Question 5

Option 4 is correct.

Let us go through the options, one by one:

1. 1x=3: incorrect, as a variable cannot start with a number
2. x 3=5: incorrect, as a variable name cannot contain a space
3. x : incorrect, as a variable needs an initial value
4. x_3=5: correct; underscore is an acceptable character while defining variables
5. x$4=5: incorrect; special characters like $ are not permissible

Question 6

Both the strings have the same value and memory address.

Output:

```
True
```

Question 7

This problem can be solved in one line – simply split the string, convert it to a list, and join it back into a string.

CODE:

```
"".join(list("123-456-7890".split("-")))
```

Question 8

Solution:

```
#Question 8
def filefunction(name):
    #open the file for writing
    with open(name+".txt","w") as f:
        f.write("Hello World")
    #read and print the file contents
    with open(name+".txt","r") as f:
        print(f.read())
    #open the file again the append mode
    with open(name+".txt","a") as f:
        f.write("\nThis is the next line")
    #reread and print the lines in the file
    with open(name+".txt","r") as f:
        print(f.read())
filename=input("Enter the name of the file ")
filefunction(filename)
```

CHAPTER 2

Exploring Containers, Classes, and Objects

In this chapter, we progress to some other essential concepts in Python - various types of containers, the methods that can be used with each of these containers, object-oriented programming, classes, and objects.

Containers

In the previous chapter, we saw that a variable could have a data type like *int, float, str,* and so on, but holds only a single value. Containers are objects that can contain multiple values. These values can have the same data type or different data types. The four built-in containers in Python are:

- Lists
- Tuples
- Dictionaries
- Sets

Containers are also called iterables; that is, you can visit or traverse through each of the values in a given container object.

In the following sections, we discuss each container and its methods in more detail.

Lists

A list contains a set of values that are stored in sequential order. A list is mutable, that is, one can modify, add, or delete values in a list, making it a flexible container.

G. Rajagopalan, *A Python Data Analyst's Toolkit*, https://doi.org/10.1007/978-1-4842-6399-0_2

An individual list item can be accessed using an index, which is an integer mentioned in square brackets, indicating the position of the item. The indexing for a list starts from 0.

Let us have a look at how to define and manipulate a list now.

Defining a list

A list can be defined by giving it a name and a set of values that are enclosed within square brackets.

CODE:

```
colors=['violet','indigo','red','blue','green','yellow']
```

Adding items to a list

Different methods can be used to add values to a list, explained in Table 2-1. The "colors" list created in the preceding code is used in the examples given in the below table.

Table 2-1. *Adding Items to a List*

Method	Description	Example

Append	Adds one item at the end of a list. The method takes only a single value as an argument.	CODE:
		`colors.append('white')`
		`#the value 'white' is added after the last item in the 'colors' list`

Insert	Adds one item at a given location or index. This method takes two arguments - the index and the value to be added.	CODE:
		`colors.insert(3,'pink')`
		`#the value 'pink' is added at the fourth position in the 'colors' list`

Extend	Adds multiple elements (as a list) at the end of a given list. This method takes another list as an argument.	CODE:
		`colors.extend(['purple','magenta'])`
		`#values 'purple' and 'magenta' added at the end of the 'colors' list`

Removing elements from a list

Just as there are multiple ways of adding an item to a list, there is more than one way to remove values from a list, as explained in Table 2-2. Note that each of these methods can remove only a single item at a time.

Table 2-2. *Removing Items from a List*

Method	Description	Example

Del	The *del* keyword deletes an item at a given location.	CODE:
		`del colors[1]`
		`#removes the second item of the 'colors' list`

Remove	This method is used when the name of the item to be removed is known, but not its position.	CODE:
		`colors.remove('white')`
		`#removes the item 'white' from the 'colors' list`

Pop	This method removes and returns the last item in the list.	CODE:
		`colors.pop()`
		`#removes the last item and displays the item removed`

Finding the index (location) of an object in the list

The *index* method is used to find out the location (or index) of a specific item or value in a list, as shown in the following statement.

CODE:

```
colors.index('violet')
```

Output:

0

Calculating the length of a list

The *len* function returns the count of the number of items in a list, as shown in the following. The name of the list is passed as an argument to this function. Note that *len* is a function, not a method. A method can be used only with an object.

CODE:

```
len(colors)
```

Output:

```
7
```

Sorting a list

The *sort* method sorts the values in the list, in ascending or descending order. By default, this method sorts the items in the ascending order. If the list contains strings, the sorting is done in alphabetical order (using the first letter of each string), as shown in the following.

CODE:

```
colors.sort()
colors
```

Output:

```
['blue', 'green', 'purple', 'red', 'violet', 'white', 'yellow']
```

Note that the list must be homogeneous (all values in the list should be of the same data type) for the sort method to work. If the list contains items of different data types, applying the sort method leads to an error.

If you want to sort your elements in the reverse alphabetical order, you need to add the *reverse* parameter and set it to "True", as shown in the following.

CODE:

```
colors.sort(reverse=True)
colors
```

Output:

```
['yellow', 'white', 'violet', 'red', 'purple', 'green', 'blue']
```

Note that if you want to just reverse the order of the items in a list, without sorting the items, you can use the *reverse* method, as shown in the following.

CODE:

```
colors=['violet','indigo','red','blue','green','yellow']
colors.reverse()
colors
```

Output:

```
['yellow', 'green', 'blue', 'red', 'indigo', 'violet']
```

Further reading:

See more on the methods that can be used with lists:

https://docs.python.org/3/tutorial/datastructures.html#more-on-lists

Slicing a list

When we create a slice from a list, we create a subset of the original list by choosing specific values from the list, based on their position or by using a condition. Slicing of a list works similar to slicing a string, which we studied in the previous chapter.

To create a slice using an index, we use the colon operator (:) and specify the start and stop values for the indexes that need to be selected.

If we provide no start or stop value before and after the colon, it is assumed that the start value is the index of the first element (0), and the stop value is the index of the last element, as shown in the following statement.

CODE:

```
`colors[:]
```

Output:

```
['Violet', 'Indigo', 'Blue', 'Green', 'Yellow', 'Orange', 'Red']
```

We can also use the colon operator twice if we are using a step index. In the following statement, alternate elements of the list are extracted by specifying a step index of two.

CODE:

```
colors[::2]
```

Output:

```
['Violet', 'Blue', 'Yellow', 'Red']
```

Just like strings, lists follow both positive and negative indexing. Negative indexing (starts from -1, which is the index of the last element in the list) works from right to left, while positive indexing (starts from 0, which is the index of the first element in the list) works from left to right.

An example of slicing with negative indexing is shown in the following, where we extract alternate elements starting from the last value in the list.

CODE:

```
colors[-1:-8:-2]
```

Output:

```
['Red', 'Yellow', 'Blue', 'Violet']
```

Creating new lists from existing lists

There are three methods for creating a new list from an existing list - list comprehensions, the *map* function, and the *filter* function - which are explained in the following.

1. **List comprehensions**

   List comprehensions provide a shorthand and intuitive way of creating a new list from an existing list.

   Let us understand this with an example where we create a new list ('colors1') from the list ('colors') we created earlier. This list only contains only those items from the original list that contain the letter 'e'.

   CODE:

   ```
   colors1=[color for color in colors if 'e' in color]
   colors1
   ```

   Output:

   ```
   ['violet', 'red', 'blue', 'green', 'yellow']
   ```

The structure of a list comprehension is explained in Figure 2-1. The output expression ('color') for items in the new list comes first. Next comes a for loop to iterate through the original list (note that other loops like the *while* loop are not used for iteration in a list comprehension). Finally, you can add an optional condition using the if/else construct.

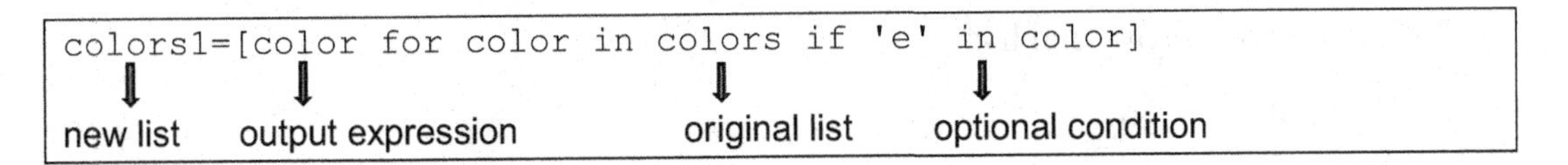

Figure 2-1. *List comprehension*

If we tried to create the same list without a list comprehension, using loops and conditions, the code would be more extended, as shown in the following.

CODE:

```
colors1=[]
for color in colors:
    if 'e' in color:
        colors1.append(color)
```

The critical point to keep in mind while using list comprehensions is that the readability of the code should not be compromised. If there are too many conditional expressions and loops involved in creating a new list, it is better to avoid list comprehensions.

Further reading: See more about list comprehensions: `https://docs.python.org/3/tutorial/datastructures.html#list-comprehensions`

2. **Map function**

   The *map* function is used to create a new list by applying a user-defined or inbuilt function on an existing list. The *map* function returns a map object, and we need to apply the list function to convert it to a list.

The map function accepts two arguments:

- The function to be applied
- The list on which the function is to be applied

In the following example, we are creating a new list ('colors1') from the 'colors' list converting its elements to uppercase. An anonymous (lambda) function is used, which is followed by the name of the original list.

CODE:

```
colors=['violet','indigo','red','blue','green','yellow']
colors1=map(lambda x:x.upper(),colors)
colors1
```

Output:

```
<map at 0x2dc87148630>
```

The function returns a map object, and the *list* function needs to be used to convert it to a list.

CODE:

```
list(colors1)
```

Output:

```
['VIOLET', 'INDIGO', 'RED', 'BLUE', 'GREEN', 'YELLOW']
```

3. **Filter function**

   The syntax of the *filter* function is similar to that of the *map* function. Whereas the *map* function returns all the objects in the original list after the function is applied, the *filter* function returns only those items that satisfy the condition specified when the filter function is called. Similar to the map function, we pass two arguments (a lambda function or a user-defined function, followed by the name of the list).

In the following example, we are creating a list from the original list, keeping only those items that have less than five characters.

CODE:

```
colors2=filter(lambda x:len(x)<5,colors)
list(colors2)
```

Output:

```
['red', 'blue']
```

Let us now discuss how we can iterate through two or more lists simultaneously.

Iterating through multiple lists using the zip function

The *zip* function provides a way of combining lists and performing operations jointly on these lists, as shown in the following. The lists that need to be combined are passed as arguments to the list function.

CODE:

```
#zip function and lists
colors=['Violet','Indigo','Blue','Green','Yellow','Orange','Red']
color_ids=[1,2,3,4,5,6,7]
for i,j in zip(colors, color_ids):
    print(i,"has a serial number",j)
```

Output:

```
Violet has a serial number 1
Indigo has a serial number 2
Blue has a serial number 3
Green has a serial number 4
Yellow has a serial number 5
Orange has a serial number 6
Red has a serial number 7
```

The *zip* function returns a list of tuples that are stored in an object of type "zip". The type of this object needs to be changed to the *list* type to view the tuples.

CODE:

```
list(zip(colors,color_ids))
```

Output:

```
[('Violet', 1),
 ('Indigo', 2),
 ('Blue', 3),
 ('Green', 4),
 ('Yellow', 5),
 ('Orange', 6),
 ('Red', 7)]
```

The next function, *enumerate*, helps us access the indexes of the items in the list.

Accessing the index of items in a list

The *enumerate* function is useful when you want to access the object as well as its index in a given list. This function returns a series of tuples, with each tuple containing the item and its index. An example of the usage of this function is shown in the following.

CODE:

```
colors=['Violet','Indigo','Blue','Green','Yellow','Orange','Red']
for index,item in enumerate(colors):
    print(item,"occurs at index",index)
```

Output:

```
Violet occurs at index 0
Indigo occurs at index 1
Blue occurs at index 2
Green occurs at index 3
Yellow occurs at index 4
Orange occurs at index 5
Red occurs at index 6
```

Concatenating of lists

The concatenation of lists, where we combine two or more lists, can be done using the '+' operator.

CODE:

```
x=[1,2,3]
y=[3,4,5]
x+y
```

Output:

```
[1, 2, 3, 3, 4, 5]
```

We can concatenate any number of lists. Note that concatenation does not modify any of the lists being joined. The result of the concatenation operation is stored in a new object.

The *extend* method can also be used for the concatenation of lists. Unlike the '+' operator, the *extend* method changes the original list.

CODE:

```
x.extend(y)
x
```

Output:

```
[1, 2, 3, 3, 4, 5]
```

Other arithmetic operators, like -, *, or /, cannot be used to combine lists.

To find the difference of elements in two lists containing numbers, we use list comprehension and the *zip* function, as shown in the following.

CODE:

```
x=[1,2,3]
y=[3,4,5]
d=[i-j for i,j in zip(x,y)]
d
```

Output:

```
[-2, -2, -2]
```

Tuples

A tuple is another container in Python, and like a list, it stores items sequentially. Like the items in a list, the values in a tuple can be accessed through their indexes. There are, however, some properties of tuples that differentiate it from lists, as explained in the following.

1. **Immutability**: A tuple is immutable, which means that you cannot add, remove, or change the elements in a tuple. A list, on the other hand, permits all these operations.

2. **Syntax**: The syntax for defining a tuple uses round brackets (or parenthesis) to enclose the individual values (in comparison with the square brackets used for a list).

3. **Speed**: In terms of speed of access and retrieval of individual elements, a tuple performs better than a list.

Let us now learn how to define a tuple and the various methods that can be used with a tuple.

Defining a tuple

A tuple can be defined with or without the parenthesis, as shown in the following code.

CODE:

```
a=(1,2,3)
#can also be defined without parenthesis
b=1,2,3
#A tuple can contain just a simple element
c=1,
#Note that we need to add the comma even though there is no element
following it because we are telling the interpreter that it is a tuple.
```

Just like a list, a tuple can contain objects of any built-in data type, like floats, strings, Boolean values, and so on.

Methods used with a tuple

While tuples cannot be changed, there are a few operations that can be performed with tuples. These operations are explained in the following.

Frequency of objects in a tuple

The *count* method is used to count the number of occurrences of a given value in a tuple:

CODE:

```
x=(True,False,True,False,True)
x.count(True)
```

Output:

```
3
```

Location/Index of a tuple item

The *index* method can be used to find the location of an item in a tuple. Using the tuple created in the previous statement, let us find the occurrence of the value, "False".

CODE:

```
x.index(False)
```

Output:

```
1
```

Only the location of the first occurrence of the item is returned by the index method.

Tuple unpacking

Tuple unpacking is the process of extracting the individual values in a tuple and storing each of these items in separate variables.

CODE:

```
a,b,c,d,e=x
```

If we do not know the number of items in a tuple, we can use the "*_" symbols to unpack the elements occurring after the first element into a list, as shown in the following.

CODE:

```
a,*_=x
print(a,_)
```

Output:

```
True [False, True, False, True]
```

Length of a tuple

The length of a tuple can be calculated using the *len* function:

CODE:

```
len(x)
```

Output:

```
5
```

Slicing of a tuple

Slicing or creation of a smaller subset of values can be performed on tuples (similar to lists and strings in this respect).

An example follows.

CODE:

```
x[::-1]
```

Output:

```
(True, False, True, False, True)
```

Applications of tuples

The following are some scenarios where tuples can be used.

1. **Creating a dictionary with tuples**

   A dictionary, which we discuss in detail in the next section, is a container containing a set of items (with a key mapping to a value). Tuples can be used for defining the items while creating a dictionary.

   A dictionary item is a tuple, and a dictionary can be defined as a list of tuples using the *dict* method, as shown in the following.

CODE:

```
x=dict([('color','pink'),('flower','rose')])
x
```

Output:

```
{'color': 'pink', 'flower': 'rose'}
```

2. **Multiple assignments**

   Tuple unpacking, as discussed earlier, is the process of separating a tuple into its components. This principle can be used for initializing multiple variables in one line, as shown in the following.

   CODE:

```
#tuple unpacking
(a,b,c,d)=range(4)
print(a,b,c,d)
```

   Output:

```
0 1 2 3
```

Further reading: See more about tuples here: https://docs.python.org/3/tutorial/datastructures.html#tuples-and-sequences

Dictionaries

A dictionary is a container that contains a set of items, with each item mapping a "key" to a "value". Each individual item is also called a key/value pair. Some other points to note about a dictionary are:

- Unlike the values in lists and tuples, the items in a dictionary are not stored in sequential order.
- Dictionaries are mutable, like lists (i.e., one can make modifications to a dictionary object).
- Curly braces are used to enclose the items in a dictionary.

Let us understand how to define a dictionary and the different methods that can be used with a dictionary object.

Defining a dictionary

A dictionary is defined as a set of comma-separated key/value pairs enclosed within a pair of curly braces, as shown in the following code.

CODE:

```
numbers={'English':'One','Spanish':'Uno','German':'Ein'}
numbers
```

Output:

```
{'English': 'One', 'Spanish': 'Uno', 'German': 'Ein'}
```

"English", "Spanish", and "German" form the keys, while "One", "Uno", and "Ein" are the values in the dictionary.

A dictionary can also be defined using the *dict* function, as explained earlier when we discussed tuples. The argument to this function is a list of tuples, with each tuple representing a key/value pair, as shown in the following.

```
numbers=dict([('English','One'),('Spanish','Uno'),('German','Ein')])
```

Adding an item (key/value pair) to a dictionary

Using the key as an index, a new item can be added to a dictionary, as shown in the following.

CODE:

```
numbers['French']='un'
numbers
#A new key/value pair with the key as 'French' and value as 'un' has been
added.
```

Output:

```
{'English': 'One', 'Spanish': 'Uno', 'German': 'Ein', 'French': 'un'}
```

Accessing the keys in a dictionary

The *keys* method to access the keys in a dictionary:

CODE:

```
numbers.keys()
```

Output:

```
dict_keys(['English', 'Spanish', 'German', 'French'])
```

Access the values in a dictionary

The *values* method to access the values in a dictionary:

CODE:

```
numbers.values()
```

Output:

```
dict_values(['One', 'Uno', 'Ein', 'un'])
```

Access all the key/value pairs in a dictionary

The *items* method is used to access the list of key/value pairs in a dictionary.

CODE:

```
numbers.items()
```

Output:

```
dict_items([('English', 'One'), ('Spanish', 'Uno'), ('German', 'Ein'),
('French', 'un')])
```

Accessing individual values

The value corresponding to a given key can be retrieved using the key as an index, as shown in the following.

CODE:

```
numbers['German']
```

Output:

```
'Ein'
```

The *get* method can also be used for retrieving values. The key is passed as an argument to this method, as shown in the following.

CODE:

```
numbers.get('German')
```

The output is the same as that obtained in the previous statement.

Setting default values for keys

The *get* method discussed in the preceding can also be used to add a key/value pair and set the default value for a key. If the key/value pair is already defined, the default value is ignored. There is another method, *setdefault,* which can also be used for this purpose.

Note that the *get* method does not change the dictionary object, while the *setdefault* method ensures that the changes are reflected in the object.

An example of the usage of the *setdefault* method is shown in the following.

CODE:

```
numbers.setdefault('Mandarin','yi')
numbers
```

Output:

```
{'English': 'One',
 'Spanish': 'Uno',
 'German': 'Ein',
 'French': 'un',
 'Mandarin': 'yi'}
```

As we can see, a new key/value pair is added.

An example of the *get* method is shown in the following.

CODE:

```
numbers.get('Hindi','Ek')
numbers
```

Output:

```
{'English': 'One',
 'Spanish': 'Uno',
```

```
 'German': 'Ein',
 'French': 'un',
 'Mandarin': 'yi'}
```

The value set by the *get* method is not added to the dictionary.

Clearing a dictionary

The *clear* method removes all the key/value pairs from a dictionary, in other words, it clears the contents of the dictionary without removing the variable from the memory.

```
#removing all the key/value pairs
numbers.clear()
```

Output:

```
{}
```

Further reading: See more about dictionaries:
https://docs.python.org/3/tutorial/datastructures.html#dictionaries

Sets

A set is a container that contains elements that are not ordered or indexed. The primary characteristic of a set is that it is a collection of *unique* elements. Note that Python does not throw an error if we add duplicate elements while creating a set. However, while we perform operations on sets, all the duplicate elements are ignored, and only distinct elements are considered.

Set definition

Just like a dictionary, a set is declared using curly braces and has unordered elements. However, while the values in a dictionary can be accessed using the keys as indexes, the values in a set cannot be accessed through indexes.

The following is an example of a set definition:

CODE:

```
a={1,1,2}
a
```

Output:

```
{1, 2}
```

As we can see from the output, the duplicate value of 1 (which is present in the set definition) is ignored.

Set operations

The methods and functions that can be used with sets are explained in Table 2-3.

Table 2-3. *Set Operations*

Operation	Method/Function	Example
Finding the length of a set	The *len* function counts the number of elements in a set, considering only the distinct values.	`len(a)`
Set iteration	The for loop can iterate through a set and print its elements.	`for x in a:` `print(x)`

Adding items or values	A single item can be added to a set using the *add* method. For adding multiple values, the *update* method is used.	`a.add(3)`
		`#or`
		`a.update([4,5])`

Removing items	Items can be removed using either the *remove* or the *discard* method.	`a.remove(4)`
		`# Or`
		`a.discard(4)`
		`#Note: When we try to delete an element that is not in the set, the discard method does not give an error, whereas the remove method gives a KeyError.`

Further reading: See more about sets: `https://docs.python.org/3/tutorial/datastructures.html#sets`

Now that we have covered all the essentials of the Python language - the concepts we learned in the previous chapter and what we understood in this chapter about the various containers and their methods, we need to decide which style or paradigm of programming to use. Among the various programming paradigms, which include procedural, functional, and object-oriented programming, we discuss object-oriented programming in the next section.

Object-oriented programming

Object-oriented programming (also commonly called "OOPS") emerged as an alternative to procedural programming, which was the traditional programming methodology.

Procedural programming involved sequential execution of a program using a series of steps. One major drawback of procedural programming is the presence of global variables that are vulnerable to accidental manipulation. OOPS offers several advantages vis-à-vis procedural programming, including the ability to reuse code, doing away with global variables, preventing unauthorized access to data, and providing the ability to manage code complexity.

Python follows the object-oriented paradigm. Classes and objects form the building blocks of object-oriented programming. Classes provide the blueprint or structure, while objects implement this structure. Classes are defined using the *class* keyword.

As an example, say you have a class named "Baby" with attributes as the name of the baby, its gender, and weight. The methods (or the functions defined within a class) for this class could be the actions performed by a baby like smiling, crying, and eating. An instance/object is an implementation of the class and has its own set of attributes and methods. In this example, each baby would have its unique characteristics (data) and behavior (functionality)

A class can have a set of attributes or variables, which may be either class variables or instance variables. All instances of the class share class variables, whereas instance variables are unique to each instance.

Let us see how classes are defined in Python, using the following example:

CODE:

```
#example of a class
class Rectangle:
    sides=4
    def __init__(self,l,b):
        self.length=l
        self.breadth=b
    def area(self):
        print("Area:",self.length*self.breadth)
my_rectangle=Rectangle(4,5)
my_rectangle.area()
```

Output:

```
Area: 20
```

The *class* keyword is followed by the name of a class and a colon sign. Following this, we are defining a class variable named "sides", and initializing it to 4. This variable is common to all objects of the class.

After this, there is a constructor function that sets or initializes the variables. Note the special syntax of the constructor function - a space follows the *def* keyword and then two underscore signs, the *init* keyword, again followed by two underscore signs.

The first parameter of any method defined in a class is the *self* keyword, which refers to an instance of the class. Then come the initialization parameters, "l" and "b", that refer to the length and breadth of the rectangle. These values are provided as arguments when we create the object. The instance variables, "self.length" and "self.breadth", are initialized using the parameters mentioned earlier. This is followed by another method that calculates the area of the rectangle. Remember that we need to add *self* as a parameter whenever we define any method of a class.

Once the class is defined, we can define an instance of this class, also called an object. We create an object just like we would create a variable, give it a name, and initialize it. "my_rectangle" is the name of the object/instance created, followed by an '=' sign.

We then mention the name of the class and the parameters used in the constructor function to initialize the object. We are creating a rectangle with length as 4 and breadth as 5. We then call the area method to calculate the area, which calculates and prints the area.

Further reading: See more about classes in Python: `https://docs.python.org/3/tutorial/classes.html`

Object-oriented programming principles

The main principles of object-oriented programming are encapsulation, polymorphism, data abstraction, and inheritance. Let us look at each of these concepts.

Encapsulation: Encapsulation refers to binding data (variables defined within a class) with the functionality (methods) that can modify it. Encapsulation also includes data hiding, since the data defined within the class is safe from manipulation by any function defined outside the class. Once we create an object of the class, its variables can be accessed and modified only by the methods (or functions) associated with the object.

Let us consider the following example:

CODE:

```
class Circle():
    def __init__(self,r):
        self.radius=r
    def area(self):
        return 3.14*self.r*self.r
c=Circle(5)
c.radius #correct way of accessing instance variable
```

Here, the class Circle has an instance variable, radius, and a method, area. The variable, radius, can only be accessed using an object of this class and not by any other means, as shown in the following statement.

CODE:

```
c.radius #correct way of accessing instance variable
radius #incorrect, leads to an error
```

Polymorphism

Polymorphism (one interface, many forms) provides the ability to use the same interface (method or function) regardless of the data type.

Let us understand the principle of polymorphism using the *len* function.

CODE:

```
#using the len function with a string
len("Hello")
```

Output:

```
5
```

CODE:

```
#using the len function with a list
len([1,2,3,4])
```

Output:

```
4
```

CODE:

```
#using the len function with a tuple
len((1,2,3))
```

Output:

```
3
```

CODE:

```
#using the len function with a dictionary
len({'a':1,'b':2})
```

Output:

```
2
```

The *len* function, which calculates the length of its argument, can take any type of argument. We passed a string, list, tuple, and dictionary as arguments to this function, and the function returned the length of each of these objects. There was no need to write a separate function for each data type.

Inheritance: Inheritance refers to the ability to create another class, called a child class, from its parent class. A child class inherits some attributes and functions from the parent class but may also have its own functionality and variables.

An example of inheritance in Python is demonstrated in the following.

```
#inheritance
class Mother():
    def __init__(self,fname,sname):
        self.firstname=fname
        self.surname=sname
    def nameprint(self):
        print("Name:",self.firstname+" "+self.surname)
class Child(Mother):
    pass
```

The parent class is called "Mother", and its attributes "firstname" and "surname" are initialized using the *init* constructor method. The child class, named "Child", is inherited from the "Mother" class. The name of the parent class is passed as an argument when we define the child class. The keyword *pass* instructs Python that nothing needs to be done for the child class (this class just inherits everything from the parent class without adding anything).

However, even if the child class does not implement any other method or add any extra attribute, the keyword *pass* is essential to prevent any error from being thrown.

Further reading: Learn more about inheritance: `https://docs.python.org/3/tutorial/classes.html#inheritance`

Data abstraction

Data abstraction is the process of presenting only the functionality while hiding the implementation details. For instance, a new user to Whatsapp needs to only learn its essential functions like sending messages, attaching photos, and placing calls, and not how these features were implemented by the developers who wrote the code for this app.

In the following example, where we declare an object of the "Circle" class and calculate the area using the area method, we do not need to know how the area is being calculated when we call the area method.

```
class Circle():
    def __init__(self,r):
        self.r=r
    def area(self):
        return 3.14*self.r*self.r
circle1=Circle(3)
circle1.area()
```

Output:

```
28.259999999999998
```

Summary

- A container is a collection of objects that belong to basic data types like int, float, str. There are four inbuilt containers in Python – lists, tuples, dictionaries, and sets.
- Each container has different properties, and a variety of functions that can be applied to Containers differ from each other depending on whether the elements can be ordered and changed (mutability) or not. Lists are mutable and ordered, tuples are immutable and ordered, and dictionaries and sets are mutable and unordered.
- Python follows the principles of object-oriented programming like inheritance (deriving a class from another class), data abstraction (presenting only the relevant detail), encapsulation (binding data with functionality), and polymorphism (ability to use an interface with multiple data types).
- A class contains a constructor function (which is defined using a special syntax), instance variables, and methods that operate on these variables. All methods must contain the keyword *self* as a parameter that refers to an object of the class.

In the next chapter, we will learn how Python can be applied in regular expressions and for solving problems in mathematics, and the libraries used for these applications.

Review Exercises

Question 1

How do you convert a list to a tuple and vice versa?

Question 2

Just like a list comprehension, a dictionary comprehension is a shortcut to create a dictionary from existing iterables. Use dictionary comprehension to create the following dictionary (from two lists, one containing the keys (a–f) and the other containing the values (1–6)):

```
{'a': 0, 'b': 1, 'c': 2, 'd': 3, 'e': 4, 'f': 5}
```

Question 3

Which of the following code statements does *not* lead to an error?

a) `'abc'[0]='d'`

b) `list('abc')[0]='d'`

c) `tuple('abc')[0]='d'`

d) `dict([('a',1),('b',2)])[0]=3`

Question 4

Write a program to calculate the number of vowels in the sentence "Every cloud has a silver lining".

Question 5

What is the output of the following code?

```
x=1,2
y=1,
z=(1,2,3)
type(x)==type(y)==type(z)
```

Question 6

What is the output of the following code?

```
numbers={
    'English':{'1':'One','2':'Two'},
    'Spanish':{'1':'Uno','2':'Dos'},
    'German':{'1':'Ein','2':'Zwei'}
}
numbers['Spanish']['2']
```

Question 7

Consider the following dictionary:

```
eatables={'chocolate':2,'ice cream':3}
```

Add another item (with "biscuit" as the key and value as 4) to this dictionary using the

- If statement
- setdefault method

Question 8

Create a list that contains odd numbers from 1 to 20 and use the appropriate list method to perform the following operations:

- Add the element 21 at the end
- insert the number 23 at the 4th position
- To this list, add another list containing even numbers from 1 to 20
- Find the index of the number 15
- Remove and return the last element
- Delete the 10th element
- Filter this list to create a new list with all numbers less than or equal to 13
- Use the map function to create a new list containing squares of the numbers in the list

- Use list comprehension to create a new list from the existing one. This list should contain the original number if it is odd. Otherwise it should contain half of that number.

Answers

Question 1

Use the list method to convert a tuple to a list:

```
list((1,2,3))
```

Use the tuple method to convert a list to a tuple:

```
tuple([1,2,3])
```

Question 2

CODE:

```
#list containing keys
l=list('abcdef')
#list containing values
m=list(range(6))
#dictionary comprehension
x={i:j for i,j in zip(l,m)}
x
```

Question 3

Correct options: b and d

In options a and c, the code statements try to change the items in a string and tuple, respectively, which are immutable objects, and hence these operations are not permitted. In options b (list) and d (dictionary), item assignment is permissible.

Question 4

Solution:

```
message="Every cloud has a silver lining"
m=message.lower()
count={}
```

```
vowels=['a','e','i','o','u']
for character in m:
    if character.casefold() in vowels:
        count.setdefault(character,0)
        count[character]=count[character]+1
print(count)
```

Question 5

Output:

```
True
```

All three methods are accepted ways of defining tuples.

Question 6

Output:

```
'Dos'
```

This question uses the concept of a nested dictionary (a dictionary within a dictionary).

Question 7

Solution:

```
eatables={'chocolate':2,'ice cream':3}
#If statement
if 'biscuit' not in eatables:
    eatables['biscuit']=4
#setdefault method(alternative method)
eatables.setdefault('biscuit',4)
```

Question 8

Solution:

```
odd_numbers=list(range(1,20,2))
#Add the element 21 at the end
odd_numbers.append(21)
#insert the number 23 at the 4th position
```

```
odd_numbers.insert(3,23)
#To this list, add another list containing even numbers from 1 to 20
even_numbers=list(range(2,20,2))
odd_numbers.extend(even_numbers)
#find the index of the number 15
odd_numbers.index(15)
#remove and return the last element
odd_numbers.pop()
#delete the 10the element
del odd_numbers[9]
#filter this list with all numbers less than or equal to 13
nos_less_13=filter(lambda x:x<=13,odd_numbers)
list(nos_less_13)
#use the map function to create a list containing squares
squared_list=map(lambda x:x**2,odd_numbers)
#use list comprehension for the new list
new_list=[x/2 if x%2==0 else x for x in odd_numbers]
new_list
```

CHAPTER 3

Regular Expressions and Math with Python

In this chapter, we discuss two modules in Python: *re*, which contains functions that can be applied for regular expressions, and *SymPy*, for solving mathematical problems in algebra, calculus, probability, and set theory. Concepts that we will learn in this chapter, like searching and replacing strings, probability, and plotting graphs, will come in handy for subsequent chapters, where we cover data analysis and statistics.

Regular expressions

A regular expression is a pattern containing both characters (like letters and digits) and metacharacters (like the * and $ symbols). Regular expressions can be used whenever we want to search, replace, or extract data with an identifiable pattern, for example, dates, postal codes, HTML tags, phone numbers, and so on. They can also be used to validate fields like passwords and email addresses, by ensuring that the input from the user is in the correct format.

Steps for solving problems with regular expressions

Support for regular expressions is provided by the *re* module in Python, which can be imported using the following statement:

```
import re
```

If you have not already installed the *re* module, go to the Anaconda Prompt and enter the following command:

```
pip install re
```


G. Rajagopalan, *A Python Data Analyst's Toolkit*, https://doi.org/10.1007/978-1-4842-6399-0_3

Once the module is imported, you need to follow the following steps.

1. **Define and compile the regular expression**: After the re module is imported, we define the regular expression and compile it. The search pattern begins with the prefix “r” followed by the string (search pattern). The “r” prefix, which stands for a raw string, tells the compiler that special characters are to be treated literally and not as escape sequences. Note that this “r” prefix is optional. The compile function compiles the search pattern into a byte code as follows and the search string (and) is passed as an argument to the compile function.

   CODE:

   ```
   search_pattern=re.compile(r'and')
   ```

2. **Locate the search pattern (regular expression) in your string**:

   In the second step, we try to locate this pattern in the string to be searched using the search method. This method is called on the variable (search_pattern) we defined in the previous step.

   CODE:

   ```
   search_pattern.search('Today and tomorrow')
   ```

   Output:

   ```
   <re.Match object; span=(6, 9), match="and">
   ```

A match object is returned since the search pattern (“and”) is found in the string (“Today and tomorrow”).

Shortcut (combining steps 2 and 3)

The preceding two steps can be combined into a single step, as shown in the following statement:

CODE:

```
re.search('and','Today and tomorrow')
```

Using one line of code, as defined previously, we combine the three steps of defining, compiling, and locating the search pattern in one step.

Further reading: Refer to this document for an introduction to using regular expressions with Python:

https://docs.python.org/3/howto/regex.html#regex-howto

Python functions for regular expressions

We use regular expressions for matching, splitting, and replacing text, and there is a separate function for each of these tasks. Table 3-1 provides a list of all these functions, along with examples of their usage.

Table 3-1. *Functions for Working with Regular Expressions in Python*

Python function	Example

re.findall( ): Searches for *all* possible matches of the regular expression and returns a list of all the matches found in the string.	CODE:
	`re.findall('3','98371234')`
	Output:
	`['3', '3']`

re.search( ): Searches for a single match and returns a match object corresponding to the first match found in the string.	CODE:
	`re.search('3','98371234')`
	Output:
	`<re.Match object; span=(2, 3), match="3">`

re.match( ): This function is similar to the *re.search* function. The limitation of this function is that it returns a match object *only* if the pattern is present *at the beginning* of the string.	CODE: `re.match('3','98371234')` Since the search pattern (3) is not present at the beginning of the string, the match function does not return an object, and we do not see any output.

(continued)

Table 3-1. *(continued)*

Python function	Example

re.split( ): Splits the string at the locations where the search pattern is found in the string being searched.	CODE:
	`re.split('3','98371234')`
	Output:
	`['98', '712', '4']`
	The string is split into smaller string wherever the search pattern, “3”, is found.

re.sub( ): Substitutes the search pattern with another string or pattern.	CODE:
	`re.sub('3','three','98371234')`
	Output:
	`'98three712three4'`
	The character “3” is replaced with the string ‘three’ in the string.

Further reading:

Learn more about the functions discussed in the above table:

- Search and match function: `https://docs.python.org/3.4/library/re.html#search-vs-match`
- Split function: `https://docs.python.org/3/library/re.html#re.split`
- Sub function: `https://docs.python.org/3/library/re.html#re.sub`
- Findall function: `https://docs.python.org/3/library/re.html#re.findall`

Metacharacters

Metacharacters are characters used in regular expressions that have a special meaning. These metacharacters are explained in the following, along with examples to demonstrate their usage.

1. **Dot (.) metacharacter**

   This metacharacter matches a single character, which could be a number, alphabet, or even itself.

   In the following example, we try to match three-letter words (from the list given after the comma in the following code), starting with the two letters "ba".

   CODE:

```
re.findall("ba.","bar bat bad ba. ban")
```

   Output:

```
['bar', 'bat', 'bad', 'ba.', 'ban']
```

   Note that one of the results shown in the output, "ba.", is an instance where the . (dot) metacharacter has matched itself.

2. **Square brackets ([]) as metacharacters**

   To match any one character among a set of characters, we use square brackets ([ ]). Within these square brackets, we define a set of characters, where one of these characters must match the characters in our text.

   Let us understand this with an example. In the following example, we try to match all strings that contain the string "ash", and start with any of following characters – 'c', 'r', 'b', 'm', 'd', 'h', or 'w'.

   CODE:

```
regex=re.compile(r'[crbmdhw]ash')
regex.findall('cash rash bash mash dash hash wash crash ash')
```

   Output:

```
['cash', 'rash', 'bash', 'mash', 'dash', 'hash', 'wash', 'rash']
```

   Note that the strings "ash" and "crash" are not matched because they do not match the criterion (the string needs to start with exactly one of the characters defined within the square brackets).

3. **Question mark (?) metacharacter**

   This metacharacter is used when you need to match at most one occurrence of a character. This means that the character we are looking for could be absent in the search string or occur just once. Consider the following example, where we try to match strings starting with the characters "Austr", ending with the characters, "ia", and having zero or one occurrence of each the following characters – "a", "l", "a", "s".

   CODE:

   ```
   regex=re.compile(r'Austr[a]?[l]?[a]?[s]?ia')
   regex.findall('Austria Australia Australasia Asia')
   ```

   Output:

   ```
   ['Austria', 'Australia', 'Australasia']
   ```

   Note that the string "Asia" does not meet this criterion.

4. **Asterisk (*) metacharacter**

   This metacharacter can match zero or more occurrences of a given search pattern. In other words, the search pattern may not occur at all in the string, or it can occur any number of times.

   Let us understand this with an example, where we try to match all strings starting with the string, "abc", and followed by zero or more occurrences of the digit –"1".

   CODE:

   ```
   re.findall("abc[1]*","abc1 abc111 abc1 abc abc111111111111 abc01")
   ```

   Output:

   ```
   ['abc1', 'abc111', 'abc1', 'abc', 'abc111111111111', 'abc']
   ```

   Note that in this step, we have combined the compilation and search of the regular expression in one single step.

5. **Backslash (\) metacharacter**

   The backslash symbol is used to indicate a character class, which is a predefined set of characters. In Table 3-2, the commonly used character classes are explained.

Table 3-2. *Character Classes*

Character Class	Characters covered
\d	Matches a digit (0–9)
\D	Matches any character that is *not* a digit
\w	Matches an alphanumeric character, which could be a lowercase letter (a–z), an uppercase letter (A–Z), or a digit (0–9)
\W	Matches any character which is *not* alphanumeric
\s	Matches any whitespace character
\S	Matches any non-whitespace character

Another usage of the backslash symbol: Escaping metacharacters

As we have seen, in regular expressions, metacharacters like . and *, have special meanings. If we want to use these characters in the literal sense, we need to "escape" them by prefixing these characters with a \(backslash) sign. For example, to search for the text W.H.O, we would need to escape the . (dot) character to prevent it from being used as a regular metacharacter.

CODE:

```
regex=re.compile(r'W\.H\.O')
regex.search('W.H.O norms')
```

Output:

```
<re.Match object; span=(0, 5), match='W.H.O'>
```

6. **Plus (+) metacharacter**

   This metacharacter matches one or more occurrences of a search pattern. The following is an example where we try to match all strings that start with at least one letter.

   CODE:

```
re.findall("[a-z]+123","a123 b123 123 ab123 xyz123")
```

   Output:

```
['a123', 'b123', 'ab123', 'xyz123']
```

7. **Curly braces {} as metacharacters**

   Using the curly braces and specifying a number within these curly braces, we can specify a range or a number representing the number of repetitions of the search pattern.

   In the following example, we find out all the phone numbers in the format "xxx-xxx-xxxx" (three digits, followed by another set of three digits, and a final set of four digits, each set separated by a "-" sign).

   CODE:

```
regex=re.compile(r'[\d]{3}-[\d]{3}-[\d]{4}')
regex.findall('987-999-8888 99122222 911-911-9111')
```

   Output:

```
['987-999-8888', '911-911-9111']
```

   Only the first and third numbers in the search string (`987-999-8888, 911-911-9111`) match the pattern. The \d metacharacter represents a digit.

   If we do not have an exact figure for the number of repetitions but know the maximum and the minimum number of repetitions, we can mention the upper and lower limit within the curly braces. In the following example, we search for all strings containing a minimum of six characters and a maximum of ten characters.

CODE:

```
regex=re.compile(r'[\w]{6,10}')
regex.findall('abcd abcd1234,abc$$$$$,abcd12 abcdef')
```

Output:

```
['abcd1234', 'abcd12', 'abcdef']
```

8. **Dollar ($) metacharacter**

   This metacharacter matches a pattern if it is present at the end of the search string.

   In the following example, we use this metacharacter to check if the search string ends with a digit.

   CODE:

   ```
   re.search(r'[\d]$','aa*5')
   ```

   Output:

   ```
   <re.Match object; span=(3, 4), match="5">
   ```

   Since the string ends with a number, a match object is returned.

9. **Caret (^) metacharacter**

   The caret (^) metacharacter looks for a match at the beginning of the string.

   In the following example, we check if the search string begins with a whitespace.

   CODE:

   ```
   re.search(r'^[\s]','   a bird')
   ```

   Output:

   ```
   <re.Match object; span=(0, 1), match=' '>
   ```

Further reading: Learn more about metacharacters: `https://docs.python.org/3.4/library/re.html#regular-expression-syntax`

Let us now discuss another library, Sympy, which is used for solving a variety of math-based problems.

Using Sympy for math problems

SymPy is a library in Python that can be used for solving a wide range of mathematical problems. We initially look at how SymPy functions can be used in algebra - for solving equations and factorizing expressions. After this, we cover a few applications in set theory and calculus.

The SymPy module can be imported using the following statement.

CODE:

```
import sympy
```

If you have not already installed the *sympy* module, go to the Anaconda Prompt and enter the following command:

```
pip install sympy
```

Let us now use this module for various mathematical problems, beginning with the factorization of expressions.

Factorization of an algebraic expression

Factorization of expressions involves splitting them into simpler expressions or factors. Multiplying these factors gives us the original expression.

As an example, an algebraic expression, like $x^2 - y^2$, can be factorized as: (x-y)*(x+y).

SymPy provides us functions for factorizing expressions as well as expanding expressions.

An algebraic expression contains variables which are represented as "symbols" in SymPy. Before SymPy functions can be applied, a variable in Python must be converted into a symbol object, which is created using the *symbols* class (for defining multiple symbols) or the *Symbol* class (for defining a single symbol). We then import the *factor* and *expand* functions, and then pass the expression we need to factorize or expand as arguments to these functions, as shown in the following.

CODE:

```
#importing the symbol classes
from sympy import symbols,Symbol
#defining the symbol objects
```

```
x,y=symbols('x,y')
a=Symbol('a')
#importing the functions
from sympy import factor,expand
#factorizing an expression
factorized_expr=factor(x**2-y**2)
#expanding an expression
expanded_expr=expand((x-y)**3)
print("After factorizing x**2-y**2:",factorized_expr)
print("After expanding (x-y)**3:",expanded_expr)
```

Output:

```
After factorizing x**2-y**2: (x - y)*(x + y)
After expanding,(x-y)**3: x**3 - 3*x**2*y + 3*x*y**2 - y**3
```

Solving algebraic equations (for one variable)

An algebraic equation contains an expression, with a series of terms, equated to zero. Let us now solve the equation $x^2 - 5x + 6 = 0$, using the *solve* function in SymPy.

We import the *solve* function from the SymPy library and pass the equation we want to solve as an argument to this function, as shown in the following. The *dict* parameter produces the output in a structured format, but including this parameter is optional.

CODE:

```
#importing the solve function
from sympy import solve
exp=x**2-5*x+6
#using the solve function to solve an equation
solve(exp,dict=True)
```

Output:

```
[{x: 2}, {x: 3}]
```

Solving simultaneous equations (for two variables)

The *solve* function can also be used to solve two equations simultaneously, as shown in the following code block.

CODE:

```
from sympy import symbols,solve
x,y=symbols('x,y')
exp1=2*x-y+4
exp2=3*x+2*y-1
solve((exp1,exp2),dict=True)
```

Output:

```
[{x: -1, y: 2}]
```

Further reading: See more on the solve function:
https://docs.sympy.org/latest/modules/solvers/solvers.html#algebraic-equations

Solving expressions entered by the user

Instead of defining the expressions, we can have the user enter expressions using the *input* function. The issue is that the input entered by the user is treated as a string, and SymPy functions are unable to process such an input.

The *sympify* function can be used to convert any expression to a type that is compatible with SymPy. Note that the user must enter mathematical operators like *, **, and so on when the input is entered. For example, if the expression is 2*x+3, the user cannot skip the asterisk symbol while entering the input. If the user enters the input as 2x+3, an error would be produced. A code example has been provided in the following code block to demonstrate the *sympify* function.

CODE:

```
from sympy import sympify,solve
expn=input("Input an expression:")
symp_expn=sympify(expn)
solve(symp_expn,dict=True)
```

Output:

```
Input an expression:x**2-9
[{x: -3}, {x: 3}]
```

Solving simultaneous equations graphically

Algebraic equations can also be solved graphically. If the equations are plotted on a graph, the point of intersection of the two lines represents the solution.

The plot function from the *sympy.plotting* module can be used to plot the equations, with the two expressions being passed as arguments to this function.

CODE:

```
from sympy.plotting import plot
%matplotlib inline
plot(x+4,3*x)
solve((x+4-y,3*x-y),dict=True)
```

Output (shown in Figure 3-1).

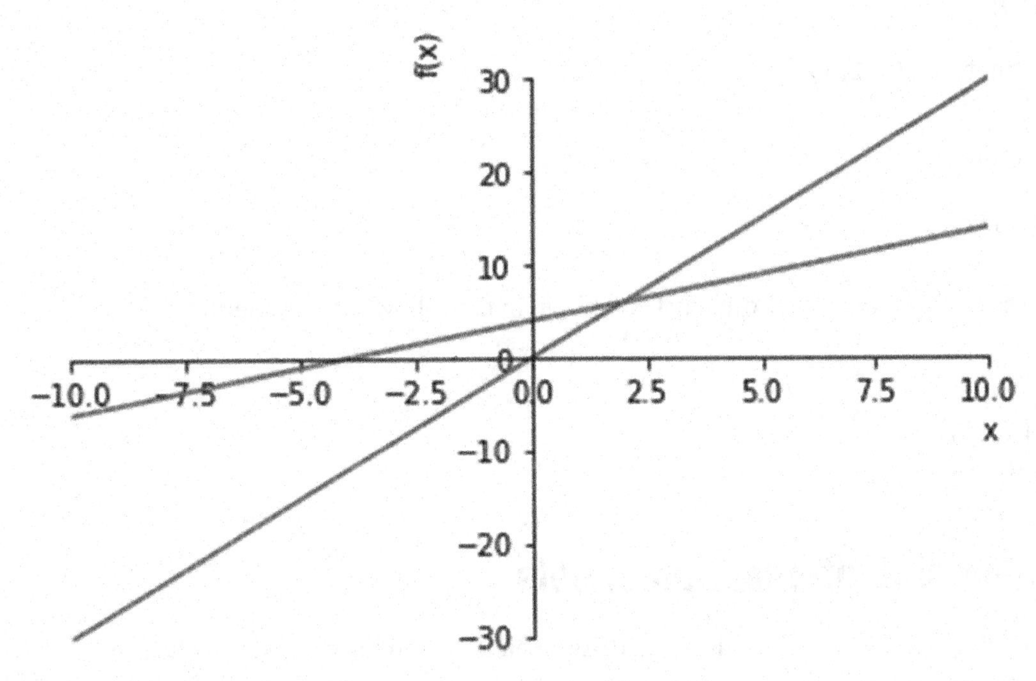

Figure 3-1. *Solving simultaneous equations with graphs*

Creating and manipulating sets

A set is a collection of unique elements, and there is a multitude of operations that can be operations that can be applied on sets. Sets are represented using Venn diagrams, which depict the relationship between two or more sets.

SymPy provides us with functions to create and manipulate sets.

First, you need to import the *FiniteSet* class from the SymPy package, to work with sets.

CODE:

```
from sympy import FiniteSet
```

Now, declare an object of this class to create a set and initialize it using the numbers you want in your set.

CODE:

```
s=FiniteSet(1,2,3)
```

Output:

```
{1,2,3}
```

We can also create a set from a list, as shown in the following statement.

CODE:

```
l=[1,2,3]
s=FiniteSet(*l)
```

Union and intersection of sets

The union of two sets is the list of all distinct elements in both the sets while the intersection of two sets includes the elements common to the two sets.

SymPy provides us a means to calculate the union and intersection of two sets using the *union* and *intersect* functions.

We use the FiniteSet class to create sets, and then apply the *union* and *intersect* functions on them, as demonstrated below.

CODE:

```
s1=FiniteSet(1,2,3)
s2=FiniteSet(2,3,4)
union_set=s1.union(s2)
intersect_set=s1.intersect(s2)
print("Union of the sets is:",union_set)
print("Intersection of the sets is:",intersect_set)
```

Output:

```
Union of the sets is: {1, 2, 3, 4}
Intersection of the sets is: {2, 3}
```

Finding the probability of an event

The probability of an event is the likelihood of the occurrence of an event, defined numerically.

Using sets to define our events and sample space, we can solve questions in probability with the help of SymPy functions.

Let us consider a simple example, where we find the probability of finding a multiple of 3 among the first ten natural numbers.

To answer this, we first define our sample space, "s", as a set with numbers from 1 to 10. Then, we define the event, denoted by the letter 'a', which is the occurrence of a multiple of 3. We then find the probability of this event ('a') by defining the number of elements in this event by the number of elements in the sample space using the *len* function. This is demonstrated in the following.

CODE:

```
s=FiniteSet(1,2,3,4,5,6,7,8,9,10)
a=FiniteSet(3,6,9)
p=len(a)/len(s)
p
```

Output:

```
0.3
```

Further reading:

See more about the operations that can be performed on sets: `https://docs.sympy.org/latest/modules/sets.html#compound-sets`

All information related to sets in SymPy: `https://docs.sympy.org/latest/modules/sets.html#module-sympy.sets.sets`

Solving questions in calculus

We will learn how to use SymPy to calculate the limiting value, derivate, and the definite and indefinite integral of a function.

Limit of a function

The limiting value of the function, f(x), is the value of the function as x approaches a particular value.

For example, if we take the function 1/x, we see that as x increases, the value of 1/x goes on reducing. As x approaches an infinitely large value, 1/x becomes closer to 0. The limiting value is calculated using the SymPy function - *limit*, as shown below.

CODE:

```
from sympy import limit,Symbol
x=Symbol('x')
limit(1/x,x,0)
```

Output:

∞

Derivative of a function

The derivative of a function defines the rate of change of this function with respect to an independent variable. If we take distance as the function and time as the independent variable, the derivate of this function is the rate of change of this function with respect to time, which is speed.

SymPy has a function, *diff*, which takes the expression (whose derivative is to be calculated) and the independent variable as arguments, and returns the derivative of the expression.

CODE:

```
from sympy import Symbol,diff
x=Symbol('x')
#defining the expression to be differentiated
expr=x**2-4
#applying the diff function to this expression
d=diff(expr,x)
d
```

Output:

$2x$

Integral of a function

The integral of a function is also called an anti-derivate. The definite integral of a function for two points, say "p" and "q", is the area under the curve between limits. For an indefinite integral, these limits are not defined.

In SymPy, the integral can be calculated using the *integrate* function.

Let us calculate the indefinite integral of the differential (2x) of the function we saw in the last example.

CODE:

```
from sympy import integrate
#applying the integrate function
integrate(d,x)
```

Output:

$x2$

Let us calculate the definite integral of the above output, using the integrate function. The arguments accepted by the integrate function include the limits, 1 and 4 (as a tuple), along with the variable (symbol), "x".

CODE:

```
integrate(d,(x,1,4))
```

Output:

```
15
```

Further reading: See more on the functions for differentiation, integration, and calculating limits: `https://docs.sympy.org/latest/tutorial/calculus.html`

Summary

1. A regular expression is a combination of literals and metacharacters and has a variety of applications.
2. A regular expression can be used to search and replace words, locating files in your system, and web crawling or scraping programs. It also has applications in data wrangling and cleaning operations, in validating the input of the user in email and HTML forms, and in search engines.
3. In Python, the *re* module provides support for regular expressions. The commonly used functions in Python for regular expression matching are: *findall*, *search*, *match*, *split*, and *sub*.
4. Metacharacters are characters with special significance in regular expressions. Each metacharacter has a specific purpose.
5. Character classes (beginning with a backslash symbol) are used to match a predefined character set, like numbers, alphanumeric characters, whitespace characters, and so on.
6. Sympy is a library used for solving mathematical problems. The basic building block used in Sympy is called a "symbol", which represents a variable. We can use the functions of the Sympy library to factorize or expand an expression, solve an equation, differentiate or integrate a function, and solve problems involving sets.

In the next chapter, we will learn another Python module, NumPy, which is used for creating arrays, computing statistical aggregation measures, and performing computations. The NumPy module also forms the backbone of Pandas, a popular library for data wrangling and analysis, which we will discuss in detail in Chapter 6.

Review Exercises

Question 1

Select the incorrect statement(s):

1. A metacharacter is considered a metacharacter even when used in sets
2. The . (dot/period) metacharacter is used to match any (single) character except a newline character
3. Regular expressions are case insensitive
4. Regular expressions, by default, return only the first match found
5. None of the above

Question 2

Explain some use cases for regular expressions.

Question 3

What is the purpose of escaping a metacharacter, and which character is used for this?

Question 4

What is the output of the following statement?

```
re.findall('bond\d{1,3}','bond07 bond007 Bond 07')
```

Question 5

Match the following metacharacters with their functions:

1. +	a. Matching zero or one character
2. *	b. Matching one or more characters
3. ?	c. Matching character sets
4. [ ]	d. Matching a character at the end of a search string
5. $	e. Matching zero or more characters
6. { }	f. Specifying an interval

Question 6

Match the following metacharacters (for character classes) with their functions:

1. \d	a. Matching the start or end of a word
2. \D	b. Matching anything other than a whitespace character
3. \S	c. Matching a nondigit
4. \w	d. Matching a digit
5. \b	e. Matching an alphanumeric character

Question 7

Write a program that asks the user to enter a password and validate it. The password should satisfy the following requirements:

- Length should be a minimum of six characters
- Contain at least one uppercase alphabet, one lowercase alphabet, one special character, and one digit

Question 8

Consider the two expressions y=x**2-9 and y=3*x-11.

Use SymPy functions to solve the following:

- Factorize the expression x**2-9, and list its factors
- Solve the two equations
- Plot the two equations and show the solution graphically
- Differentiate the expression x**2-9, for x=1
- Find the definite integral of the expression 3*x-11 between points x=0 and x=1

Answers

Question 1

The incorrect options are options 1 and 3.

Option 1 is incorrect because when a metacharacter is used in a set, it is not considered a metacharacter and assumes its literal meaning.

Option 3 is incorrect because regular expressions are case sensitive ("hat" is not the same as "HAT").

The other options are correct.

Question 2

Some use cases of regular expressions include

1. User input validation in HTML forms. Regular expressions can be used to check the user input and ensure that the input is entered as per the requirements for various fields in the form.
2. Web crawling and web scraping: Regular expressions are commonly used for searching for general information from websites (crawling) and for extracting certain kinds of text or data from websites (scraping), for example phone numbers and email addresses.
3. Locating files on your operating system: Using regular expressions, you can search for files on your system that have file names with the same extension or following some other pattern.

Question 3

We escape a metacharacter to use it in its literal sense. The backslash character (\) symbol precedes the metacharacter you want to escape. For instance, the symbol "*" has a special meaning in a regular expression. If you want to use this character it without the special meaning, you need to use *

Question 4

Output

```
['bond07', 'bond007']
```

Question 5

```
1-b; 2-e; 3-a; 4-d; 5-c; 6-f
```

Question 6

```
1-d; 2-c; 3-b; 4-e; 5-a
```

Question 7

CODE:

```
import re
special_characters=['$','#','@','&','^','*']
while True:
    s=input("Enter your password")
    if len(s)<6:
        print("Enter at least 6 characters in your password")
    else:
        if re.search(r'\d',s) is None:
            print("Your password should contain at least 1 digit")
        elif re.search(r'[A-Z]',s) is None:
            print("Your password should contain at least 1 uppercase letter")
        elif re.search(r'[a-z]',s) is None:
            print("Your password should contain at least 1 lowercase letter")
        elif not any(char in special_characters for char in s):
            print("Your password should contain at least 1 special character")
        else:
            print("The password you entered meets our requirements")
            break
```

Question 8

CODE:

```
from sympy import Symbol,symbols,factor,solve,diff,integrate,plot
#creating symbols
x,y=symbols('x,y')
y=x**2-9
y=3*x-11
```

```
#factorizing the expression
factor(x**2-9)
#solving two equations
solve((x**2-9-y,3*x-11-y),dict=True)
#plotting the equations to find the solution
%matplotlib inline
plot(x**2-9-y,3*x-11-y)
#differentiating at a particular point
diff(x**2-9,x).subs({x:1})
#finding the integral between two points
integrate(3*x-11,(x,0,1))
```

CHAPTER 4

Descriptive Data Analysis Basics

In previous chapters, you were introduced to the Python language – the syntax, functions, conditional statements, data types, and different types of containers. You also reviewed more advanced concepts like regular expressions, handling of files, and solving mathematical problems with Python. Our focus now turns to the meat of the book, descriptive data analysis (also called exploratory data analysis).

In descriptive data analysis, we analyze past data with the help of methods like summarization, aggregation, and visualization to draw meaningful insights. In contrast, when we do predictive analytics, we try to make predictions or forecasts about the future using various modeling techniques.

In this chapter, we look at the various types of data, how to classify data, which operations to perform based on the category of data, and the workflow of the descriptive data analysis process.

Descriptive data analysis - Steps

Figure 4-1 illustrates the methodology followed in descriptive data analysis, step by step.

G. Rajagopalan, *A Python Data Analyst's Toolkit*, https://doi.org/10.1007/978-1-4842-6399-0_4

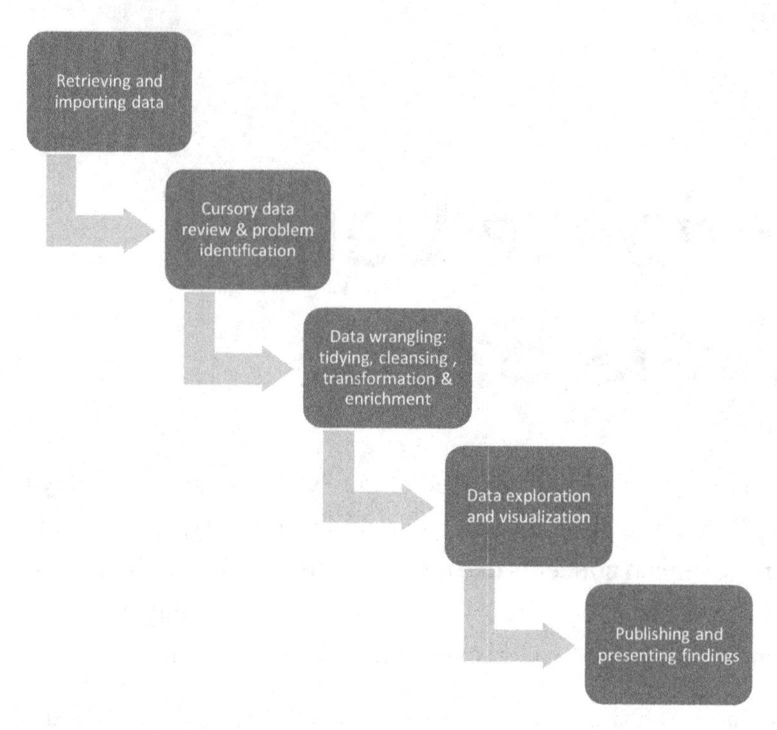

Figure 4-1. *Steps in descriptive data analysis*

Let us understand each of these steps in detail.

1) **Data retrieval**: Data could be stored in a structured format (like databases or spreadsheets) or an unstructured format (like web pages, emails, Word documents). After considering parameters such as the cost and structure of the data, we need to figure out how to retrieve this data. Libraries like Pandas provide functions for importing data in a variety of formats.

2) **Cursory data review and problem identification**: In this step, we form first impressions of the data that we want to analyze. We aim to understand each of the individual columns or features, the meanings of various abbreviations and notations used in the dataset, what the records or data represent, and the units used for the data storage. We also need to ask the right questions and figure out what we need to do before getting into the nitty-gritty of our analysis. These questions may include the following: which

are the features that are relevant for analysis, is there an increasing or decreasing trend in individual columns, do we see any missing values, are we trying to develop a forecast and predict one feature, and so on.

3) **Data wrangling**: This step is the crux of data analysis and the most time-consuming activity, with data analysts and scientists spending approximately 80% of their time on this.

   Data in its raw form is often unsuitable for analysis due to any of the following reasons: presence of missing and redundant values, outliers, incorrect data types, presence of extraneous data, more than one unit of measurement being used, data being scattered across different sources, and columns not being correctly identified.

   Data wrangling or munging is the process of transforming raw data so that it is suitable for mathematical processing and plotting graphs. It involves removing or substituting missing values and incomplete entries, getting rid of filler values like semicolons and commas, filtering the data, changing data types, eliminating redundancy, and merging data with other sources.

   Data wrangling comprises tidying, cleansing, and enriching data. In data tidying, we identify the variables in our dataset and map them to columns. We also structure data along the right axis and ensure that the rows contain observations and not features. The purpose of converting data into a tidy form is to have data in a structure that facilitates ease of analysis. Data cleansing involves dealing with missing values, incorrect data types, outliers, and wrongly entered data. In data enrichment, we may add data from other sources and create new columns or features that may be helpful for our analysis.

4) **Data exploration and visualization**: After the data has been prepared, the next step involves finding patterns in data, summarizing key characteristics, and understanding relationships among various features. With visualization, you can achieve all of this, and also lucidly present critical findings. Python libraries for visualization include Matplotlib, Seaborn, and Pandas.

5) **Presenting and publishing our analysis**: Jupyter notebooks serve the dual purpose of both executing our code and serving as a platform to provide a high-level summary of our analysis. By adding notes, headings, annotations, and images, you can spruce up your notebook to make it presentable to a broader audience. The notebook can be downloaded in a variety of formats, like PDF, which can later be shared with others for review.

We now move on to the various structures and levels of data.

Structure of data

The data that we need to analyze could have any of the following structures, demonstrated in Figure 4-2.

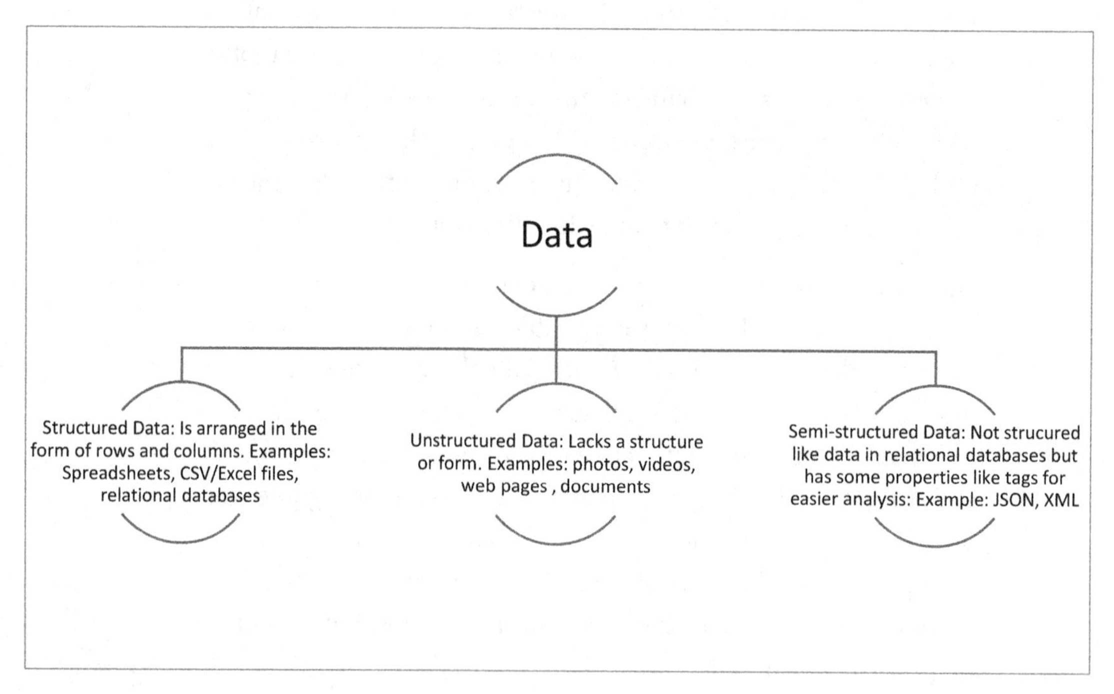

Figure 4-2. *Structure of data*

Classifying data into different levels

There are broadly two levels of data: Continuous and Categorical. Continuous data can further be classified as ratio and interval, while categorical data can be either nominal or ordinal. The levels of data are demonstrated in Figure 4-3.

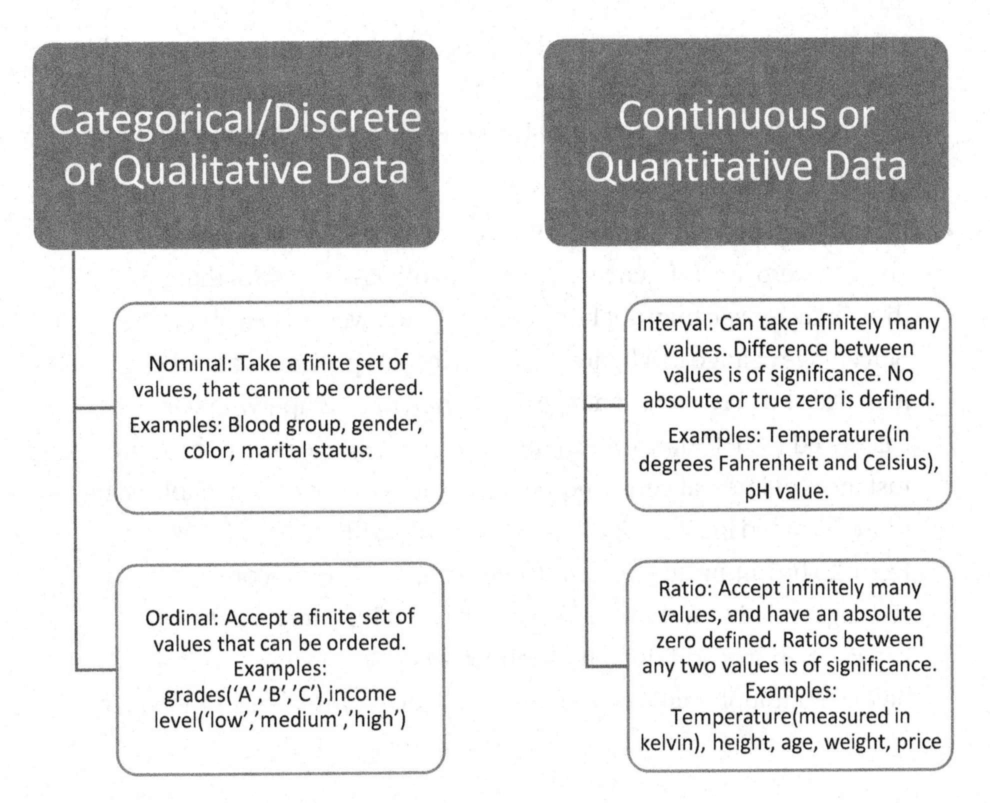

Figure 4-3. *Levels of data*

The following are some essential points to note:

- **Numeric values for categorical variables**: Categorical data is not restricted to non-numeric values. For example, the rank of a student, which could take values like 1/2/3 and so on, is an example of an ordinal (categorical) variable that contains numbers as values. However, these numbers do not have mathematical significance; for instance, it would not make sense to find the average rank.
- **Significance of a true zero point**: We have noted that interval variables do not have an absolute zero as a reference point, while ratio variables have a valid zero point. An absolute zero denotes the absence of a value. For example, when we say that variables like height and weight are ratio variables, it would mean that a value of 0 for any of these variables would mean an invalid or nonexistent data point. For an interval variable like temperature (when measured in degrees Celsius or Fahrenheit), a value of 0 does not mean that data

is absent. 0 is just one among the values that the temperature variable can assume. On the other hand, temperature, when measured in the Kelvin scale, is a ratio variable since there is an absolute zero defined for this scale.

- **Identifying interval variables**: Interval variables do not have an absolute zero as a reference point, but identifying variables that have this characteristic may not be apparent. Whenever we talk about the percentage change in a figure, it is relative to its previous value. For instance, the percentage change in inflation or unemployment is calculated with the last value in time as the reference point. These are instances of interval data. Another example of an interval variable is the score obtained in a standardized test like the GRE (Graduate Record Exam). The minimum score is 260, and the maximum score is 340. The scoring is relative and does not start from 0. With interval data, while you can perform addition and subtraction operations. You cannot divide or multiply values (operations that are permissible for ratio data).

Visualizing various levels of data

Whenever you need to analyze data, first understand if the data is structured or unstructured. If the data is unstructured, convert it to a structured form with rows and columns, which makes it easier for further analysis using libraries like Pandas. Once you have data in this format, categorize each of the features or columns into the four levels of data and perform your analysis accordingly.

Note that in this chapter, we only aim to understand how to categorize the variables in a dataset and identify the operations and plots that would apply for each category. The actual code that needs to be written to visualize the data is explained in Chapter 7.

We look at how to classify the features and perform various operations using the famous *Titanic* dataset. The dataset can be imported from here:

`https://github.com/DataRepo2019/Data-files/blob/master/titanic.csv`

Background information about the dataset: The RMS *Titanic*, a British passenger ship, sank on its maiden voyage from Southampton to New York on 15th April 1912, after it collided with an iceberg. Out of the 2,224 passengers, 1,500 died, making this event a tragedy of epic proportions. This dataset describes the survival status of the passengers and other details about them, including their class, name, age, and the number of relatives.

Figure 4-4 provides a snapshot of this dataset.

	PassengerId	Survived	Pclass	Name	Sex	Age	SibSp	Parch	Ticket	Fare	Cabin	Embarked
0	1	0	3	Braund, Mr. Owen Harris	male	22.0	1	0	A/5 21171	7.2500	NaN	S
1	2	1	1	Cumings, Mrs. John Bradley (Florence Briggs Th...	female	38.0	1	0	PC 17599	71.2833	C85	C
2	3	1	3	Heikkinen, Miss. Laina	female	26.0	0	0	STON/O2. 3101282	7.9250	NaN	S
3	4	1	1	Futrelle, Mrs. Jacques Heath (Lily May Peel)	female	35.0	1	0	113803	53.1000	C123	S
4	5	0	3	Allen, Mr. William Henry	male	35.0	0	0	373450	8.0500	NaN	S

Figure 4-4. *Titanic dataset*

The features in this dataset, classified according to the data level, are captured in Table 4-1.

Table 4-1. *Titanic Dataset – Data Levels*

Feature in the dataset	What it represents	Level of data
PassengerId	Identity number of passenger	Nominal
Pclass	Passenger class (1:1st class; 2: 2nd class; 3: 3rd class), passenger class is used as a measure of the socioeconomic status of the passenger	Ordinal
Survived	Survival status (0:Not survived; 1:Survived)	Nominal
Name	Name of passenger	Nominal
Sibsp	Number of siblings/spouses aboard	Ratio
Ticket	Ticket number	Nominal
Cabin	Cabin number	Nominal
Sex	Gender of passenger	Nominal
Age	Age	Ratio
Parch	Number of parents/children aboard	Ratio
Fare	Passenger fare (British pound)	Ratio
Embarked	Port of embarkation (with C being Cherbourg, Q being Queenstown, and S being Southampton)	Nominal

Let us now understand the rationale behind the classification of the features in this dataset.

1. Nominal variables: Variables like "PassengerId", "Survived", "Name", "Sex", "Cabin", and "Embarked" do not have any intrinsic ordering of their values. Note that some of these variables have numeric values, but these values are finite in number. We cannot perform an arithmetic operation on these values like addition, subtraction, multiplication, or division. One operation that is common with nominal variables is counting. A commonly used method in Pandas, *value_counts* (discussed in the next chapter), is used to determine the number of values per each unique category of the nominal variable. We can also find the mode (the most frequently occurring value). The bar graph is frequently used to visualize nominal data (pie charts can also be used), as shown in Figure 4-5.

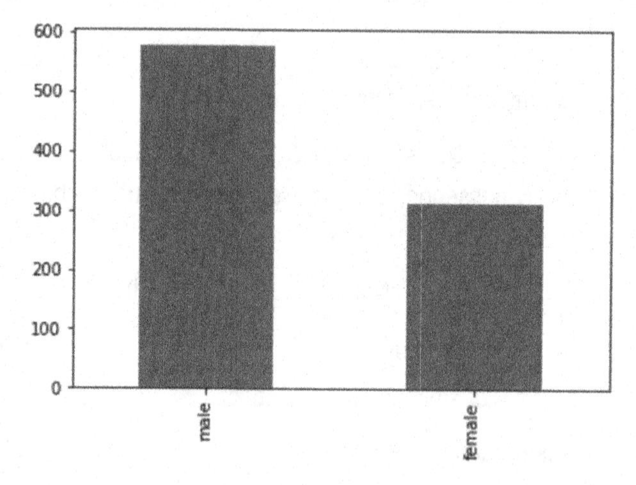

***Figure 4-5.** Bar graph showing the count of each category*

2. Ordinal variables: "Pclass" (or Passenger Class) is an ordinal variable since its values follow an order. A value of 1 is equivalent to first class, 2 is equivalent to the second class, and so on. These class values are indicative of socioeconomic status.

We can find out the median value and percentiles. We can also count the number of values in each category, calculate the mode, and use plots like bar graphs and pie charts, just as we did for nominal variables.

In Figure 4-6, we have used a pie chart for the ordinal variable "Pclass".

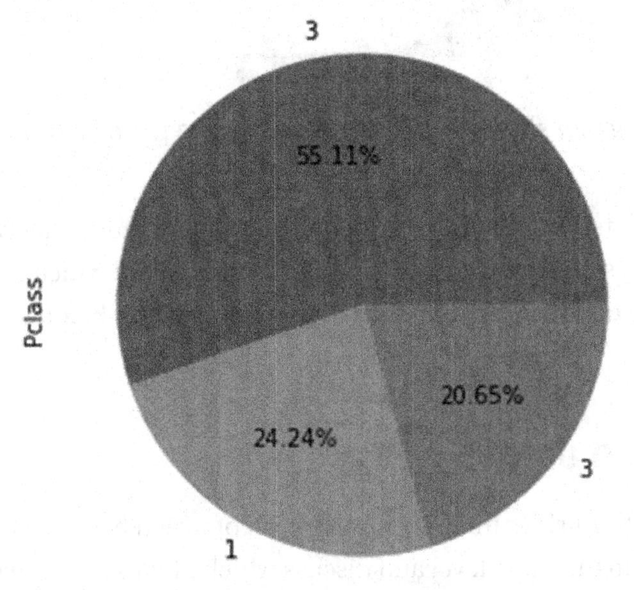

***Figure 4-6.** Pie chart showing the percentage distribution of each class*

3. Ratio Data: The "Age" and "Fare" variables are examples of ratio data, with the value zero as a reference point. With this type of data, we can perform a wide range of mathematical operations.

   For example, we can add all the fares and divide it by the total number of passengers to find the mean. We can also find out the standard deviation. A histogram, as shown in Figure 4-7, can be used to visualize this kind of continuous data to understand the distribution.

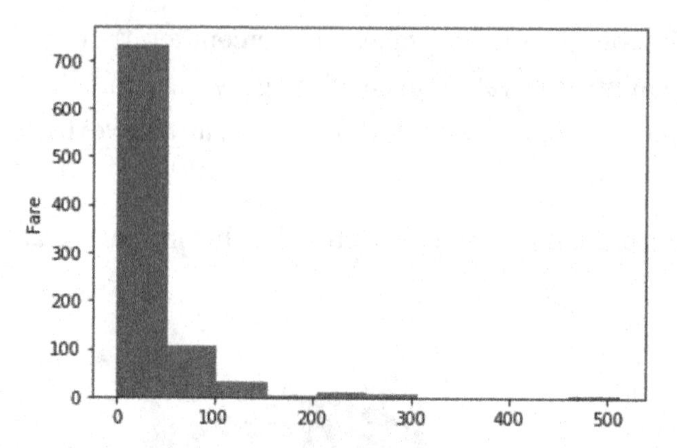

Figure 4-7. *Histogram showing the distribution of a ratio variable*

In the preceding plots, we looked at the graphs for plotting individual categorical or continuous variables. In the following section, we understand which graphs to use when we have more than one variable or a combination of variables belong to different scales or levels.

Plotting mixed data

In this section, we'll consider three scenarios, each of which has two variables that may or may not belong to the same level and discuss which plot to use for each scenario (using the same *Titanic* dataset).

1. One categorical and one continuous variable: A box plot shows the distribution, symmetry, and outliers for a continuous variable. A box plot can also show the continuous variable against a categorical variable. In Figure 4-8, the distribution of 'Age' (a ratio variable) for each value of the nominal variable – 'Survived' (0 is the value for passengers who did not survive and 1 is the value for those who did).

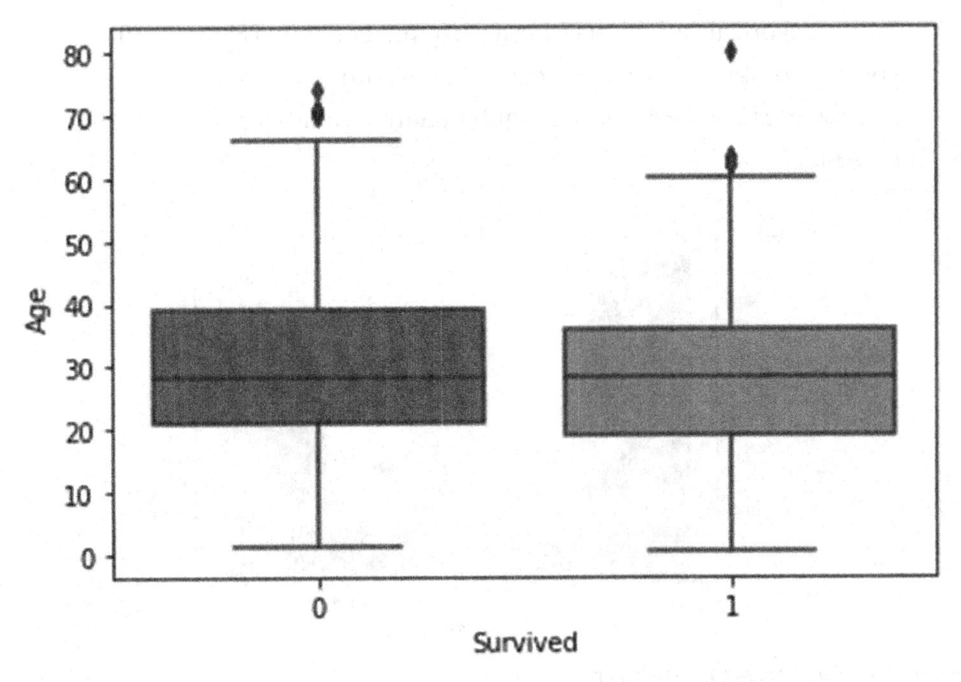

***Figure 4-8.** Box plot, showing the distribution of age for different categories*

2. Both continuous variables: Scatter plots are used to depict the relationship between two continuous variables. In Figure 4-9, we plot two ratio variables, 'Age' and 'Fare', on the x and y axes to produce a scatter plot.

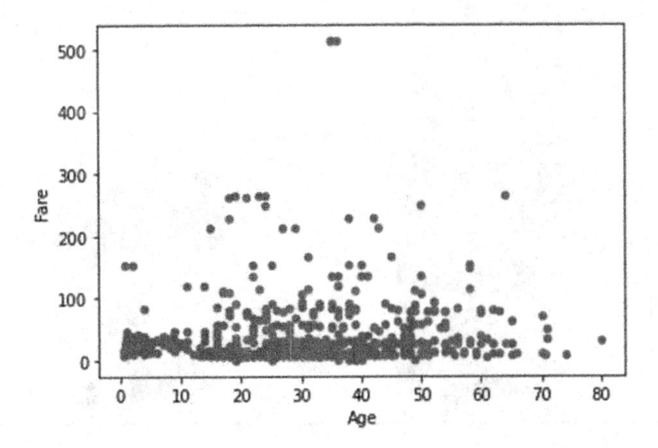

***Figure 4-9.** Scatter plot*

3. Both categorical variables: Using a clustered bar chart (Figure 4-10), you can combine two categorical variables with the bars depicted side by side to represent every combination of values for the two variables.

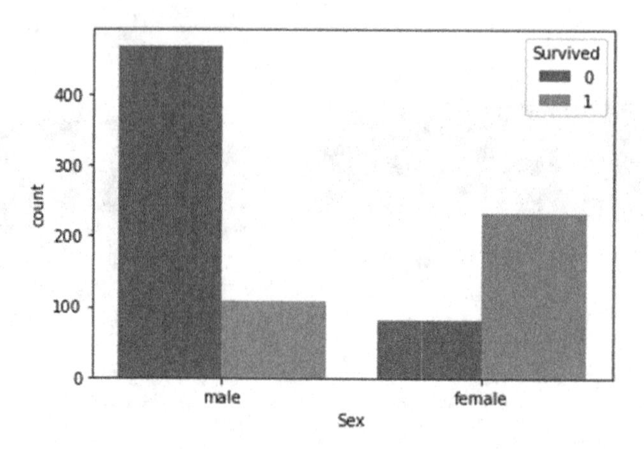

***Figure 4-10.** Clustered bar chart*

We can also use a stacked bar chart to plot two categorical variables. Consider the following stacked bar chart, shown in Figure 4-11, plotting two categorical variables - "Pclass" and "Survived".

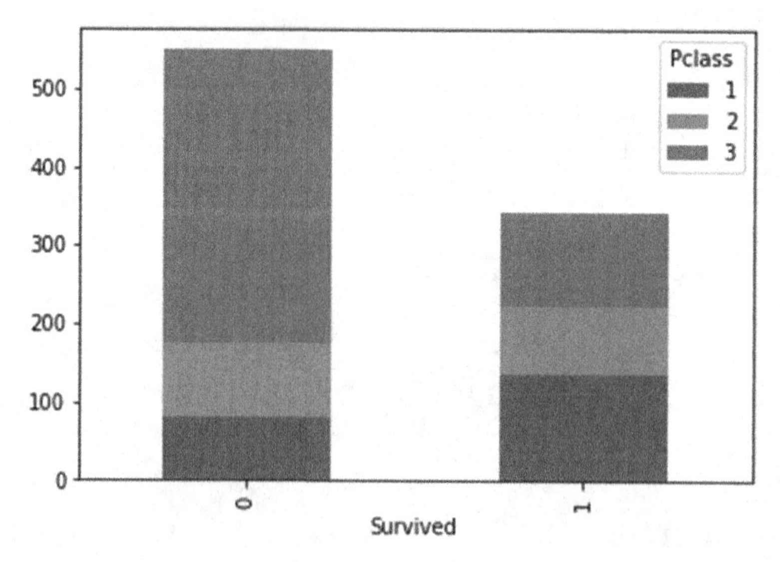

***Figure 4-11.** Stacked bar chart*

In summary, you can use a scatter plot for two continuous variables, a stacked or clustered bar chart for two categorical variables, and a box plot when you want to display a continuous variable across different values of a categorical variable.

Summary

1. Descriptive data analysis is a five-step process that uses past data and follows a stepwise methodology. The core of this process - data wrangling - involves dealing with missing values and other anomalies. It also deals with restructuring, merging, and transformations.
2. Data can be classified based on its structure (structured, unstructured, or semistructured) or based on the type of values it contains (categorical or continuous).
3. Categorical data can be classified as nominal and ordinal (depending on whether the values can be ordered or not). Continuous data can be of either ratio or interval type (depending on whether the data has 0 as an absolute reference point).
4. The kind of mathematical operations and graphical plots that can be used varies, depending on the level of data.

Now that you have gained a high-level perspective of the descriptive data analysis process, we get into the nitty-gritty of data analysis in the next chapter. We look at how to write code for various tasks that we perform in data wrangling and preparation in the following chapter that covers the Pandas library.

Review Exercises

Question 1

Classify the following variables based on the type of data.

- pH scale
- Language proficiency

- Likert Scale (used in surveys)
- Work experience
- Time of the day
- Social security number
- Distance
- Year of birth

Question 2

Arrange the following five steps in the order in which they occur during the data analysis process.

1. Visualization
2. Publishing and presentation of analysis
3. Importing data
4. Data wrangling
5. Problem statement formulation

Question 3

For each of the following operations or statistical measures, list the compatible data types.

- Division
- Addition
- Multiplication
- Subtraction
- Mean
- Median
- Mode
- Standard deviation
- Range

Question 4

For each of the following list the compatible data types.

- Bar graphs
- Histograms
- Pie charts
- Scatter plots
- Stacked bar charts

Answers

Question 1

- pH scale: Interval

  The pH scale does not have an absolute zero point. While the values can be compared, we cannot calculate ratios.

- Language proficiency: Ordinal

  Proficiency in a language has various levels like "beginner", "intermediate", and "advanced" that are ordered, and hence come under the ordinal scale.

- Likert Scale (used in surveys): Ordinal.

  The Likert Scale is often used in surveys, with values like "not satisfied", "satisfied", and "very satisfied". These values form a logical order, and therefore any variable representing the Likert Scale is an ordinal variable.

- Work experience: Ratio

  As there is an absolute zero for this variable and one can perform arithmetic operations, including calculation of ratios, this variable is a ratio variable.

- Time of the day: Interval

  Time (on a 12-hour scale) does not have an absolute zero point. We can calculate the difference between two points of time, but cannot calculate ratios.

- Social security number: Nominal

  Values for identifiers like social security numbers are not ordered and do not lend themselves to mathematical operations.

- Distance: Ratio

  With a reference point as 0 and values that can be added, subtracted, multiplied, and divided, distance is a ratio variable.

- Year of birth: Interval

  There is no absolute zero point for such a variable. You can calculate the difference between two years, but we cannot find out ratios.

Question 2

The correct order is 3, 5, 4, 1, 2

Question 3

- Division: Ratio data
- Addition: Ratio data, interval data
- Multiplication: Ratio data
- Subtraction: Interval data, ratio data
- Mean: Ratio data, interval data
- Median: Ordinal data, ratio data, interval data
- Mode: All four levels of data (ratio, interval, nominal, and ordinal)
- Standard deviation: Ratio and interval data
- Range: Ratio and interval data

Question 4

- Box plots: Ordinal, ratio, interval
- Histograms: Ratio, interval
- Pie charts: Nominal, ordinal
- Scatter plots: Ratio, interval
- Stacked bar charts: Nominal, ordinal

CHAPTER 5

Working with NumPy Arrays

NumPy, or Numerical Python, is a Python-based library for mathematical computations and processing arrays. Python does not support data structures in more than one dimension, with containers like lists, tuples, and dictionaries being unidimensional. The inbuilt data types and containers in Python cannot be restructured into more than one dimension, and also do not lend themselves to complex computations. These drawbacks are limitations for some of the tasks involved while analyzing data and building models, which makes arrays a vital data structure.

NumPy arrays can be reshaped and utilize the principle of vectorization (where an operation applied to the array reflects on all its elements).

In the previous chapter, we looked at the basic concepts used in descriptive data analysis. NumPy is an integral part of many of the tasks we perform in data analysis, serving as the backbone of many of the functions and data types used in Pandas. In this chapter, we understand how to create NumPy arrays using a variety of methods, combine arrays, and slice, reshape, and perform computations on them.

Getting familiar with arrays and NumPy functions

Here, we look at various methods of creating and combining arrays, along with commonly used NumPy functions.

Importing the NumPy package

The NumPy package has to be imported before its functions can be used, as shown in the following. The shorthand notation or alias for NumPy is *np*.

CODE:

```
import numpy as np
```

If you have not already installed NumPy, go to the Anaconda Prompt and enter the following command:

```
pip install numpy
```


G. Rajagopalan, *A Python Data Analyst's Toolkit*, https://doi.org/10.1007/978-1-4842-6399-0_5

Creating an array

The basic unit in NumPy is an array. In Table 5-1, we look at the various methods for creating an array.

Table 5-1. *Methods of Creating NumPy Arrays*

Method	Example

Creating an array from a list	The *np.array* function is used to create a one-dimensional or multidimensional array from a list.
	CODE:
	`np.array([[1,2,3],[4,5,6]])`
	Output:
	`array([[1, 2, 3],`
	`       [4, 5, 6]])`

Creating an array from a range	The *np.arange* function is used to create a range of integers.
	CODE:
	`np.arange(0,9)`
	`#Alternate syntax:`
	`np.arange(9)`
	`#Generates 9 equally spaced integers starting from 0`
	Output:
	`array([0, 1, 2, 3, 4, 5, 6, 7, 8])`

Creating an array of equally spaced numbers	The *np.linspace* function creates a given number of equally spaced values between two limits.
	CODE:
	`np.linspace(1,6,5)`
	`# This generates five equally spaced values between 1 and 6`
	Output:
	`array([1.  , 2.25, 3.5 , 4.75, 6.  ])`

(*continued*)

Table 5-1. (*continued*)

Method	Example

Creating an array of zeros	The *np.zeros* function creates an array with a given number of rows and columns, with only one value throughout the array – "0".
	CODE:
	`np.zeros((4,2))`
	`#Creates a 4*2 array with all values as 0`
	Output:
	`array([[0., 0.],`
	`       [0., 0.],`
	`       [0., 0.],`
	`       [0., 0.]])`

Creating an array of ones	The *np.ones* function is similar to the *np.zeros* function, the difference being that the value repeated throughout the array is "1".
	CODE:
	`np.ones((2,3))`
	`#creates a 2*3 array with all values as 1`
	Output:
	`array([[1., 1., 1.],`
	`       [1., 1., 1.]])`

Creating an array with a given value repeated throughout	The *np.full* function creates an array using the value specified by the user.
	CODE:
	`np.full((2,2),3)`
	`#Creates a 2*2 array with all values as 3`
	Output:
	`array([[3, 3],`
	`       [3, 3]])`

(*continued*)

Table 5-1. *(continued)*

Method	Example

Creating an empty array	The *np.empty* function generates an array, without any particular initial value (array is randomly initialized).
	CODE:
	`np.empty((2,2))`
	`#creates a 2*2 array filled with random values`
	Output:
	`array([[1.31456805e-311, 9.34839993e+025],`
	`       [2.15196058e-013, 2.00166813e-090]])`

Creating an array from a repeating list	The *np.repeat* function creates an array from a list that is repeated a given number of times.
	CODE:
	`np.repeat([1,2,3],3)`
	`#Will repeat each value in the list 3 times`
	Output:
	`array([1, 1, 1, 2, 2, 2, 3, 3, 3])`

Creating an array of random integers	The *randint* function (from the *np.random* module) generates an array containing random numbers.
	CODE:
	`np.random.randint(1,100,5)`
	`#Will generate an array with 5 random numbers between 1 and 100`
	Output:
	`array([34, 69, 67,  3, 96])`

One point to note is that arrays are homogeneous data structures, unlike containers (like lists, tuples, and dictionaries); that is, an array should contain items of the same data type. For example, we cannot have an array containing integers, strings, and floating-point (decimal) values together. While defining a NumPy array with items of different data types does not lead to an error while you write code, it should be avoided.

Now that we have looked at the various ways of defining an array, we look at the operations that we can perform on them, starting with the reshaping of an array.

Reshaping an array

Reshaping an array is the process of changing the dimensionality of an array. The NumPy method "reshape" is important and is commonly used to convert a 1-D array to a multidimensional one.

Consider a simple 1-D array containing ten elements, as shown in the following statement.

CODE:

```
x=np.arange(0,10)
```

Output:

```
array([0, 1, 2, 3, 4, 5, 6, 7, 8, 9])
```

We can reshape the 1-D array "x" into a 2-D array with five rows and two columns:

CODE:

```
x.reshape(5,2)
```

Output:

```
array([[0, 1],
       [2, 3],
       [4, 5],
       [6, 7],
       [8, 9]])
```

As another example, consider the following array:

CODE:

```
x=np.arange(0,12)
x
```

Output:

```
array([ 0,  1,  2,  3,  4,  5,  6,  7,  8,  9, 10, 11])
```

Now, apply the *reshape* method to create two subarrays - each with three rows and two columns:

CODE:

```
x=np.arange(0,12).reshape(2,3,2)
x
```

Output:

```
array([[[ 0,  1],
        [ 2,  3],
        [ 4,  5]],

       [[ 6,  7],
        [ 8,  9],
        [10, 11]]])
```

The product of the dimensions of the reshaped array should equal the number of elements in the original array. In this case, the dimensions of the array (2,3,2) when multiplied equal 12, the number of elements in the array. If this condition is not satisfied, reshaping fails to work.

Apart from the *reshape* method, we can also use the *shape* attribute to change the shape or dimensions of an array:

CODE:

```
x.shape=(5,2)
#5 is the number of rows, 2 is the number of columns
```

Note that the *shape* attribute makes changes to the original array, while the *reshape* method does not alter the array.

The reshaping process can be reversed using the "ravel" method:

CODE:

```
x=np.arange(0,12).reshape(2,3,2)
x.ravel()
```

Output:

```
array([ 0,  1,  2,  3,  4,  5,  6,  7,  8,  9, 10, 11])
```

Further reading: See more on array creation routines: `https://numpy.org/doc/stable/reference/routines.array-creation.html#`

The logical structure of arrays

The cartesian coordinate system, which is used to specify the location of a point, consists of a plane with two perpendicular lines known as the "x" and "y" axes. The position of a point is specified using its x and y coordinates. This principle of using axes to represent different dimensions is also used in arrays.

A 1-D array has one axis (axis=0) as it has one dimension, as shown in Figure 5-1.

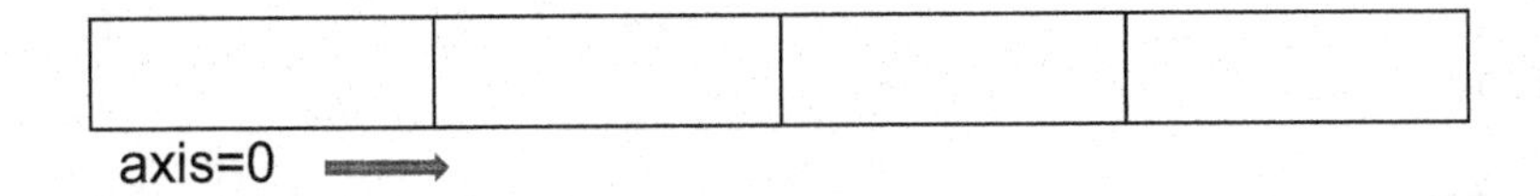

***Figure 5-1.** 1-D array representation*

A 2-D array has an axis value of "0" to represent the row axis and a value of "1" to represent the column axis, as shown in Figure 5-2.

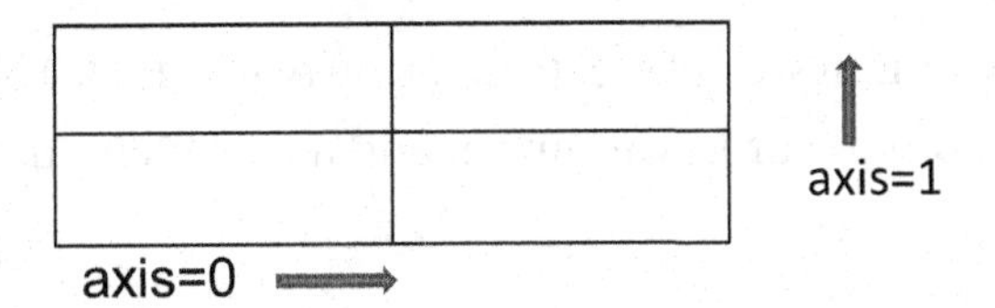

***Figure 5-2.** A 2-D array representation*

A 3-D array has three axes, representing three dimensions, as shown in Figure 5-3.

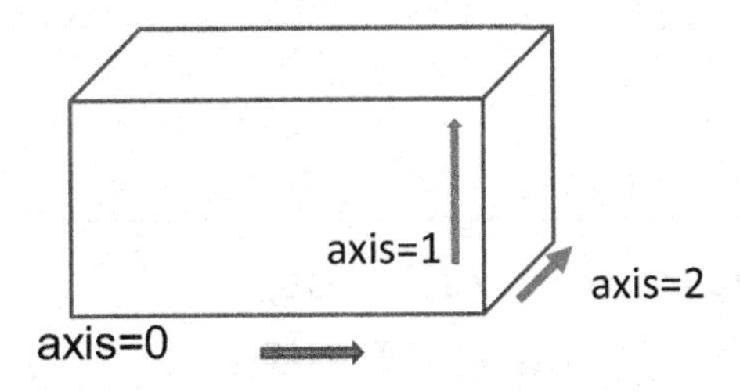

***Figure 5-3.** A 3-D array representation*

Extending the logic, an array with "n" dimensions has "n" axes.

Note that the preceding diagrams represent only the logical structure of arrays. When it comes to storage in the memory, elements in an array occupy contiguous locations, regardless of the dimensions.

Data type of a NumPy array

The *type* function can be used to determine the type of a NumPy array:

CODE:

```
type(np.array([1,2,3,4]))
```

Output:

```
numpy.ndarray
```

Modifying arrays

The length of an array is set when you define it. Let us consider the following array, "a":

CODE:

```
a=np.array([0,1,2])
```

The preceding code statement would create an array of length 3. The array length is not modifiable after this. In other words, we cannot add a new element to the array after its definition.

The following statement, where we try to add a fourth element to this array, would lead to an error:

CODE:

```
a[3]=4
```

Output:

```
---------------------------------------------------------------------------
IndexErrorTraceback (most recent call last)
<ipython-input-215-94b083a55c38> in <module>
----> 1a[3]=4

IndexError: index 3 is out of bounds for axis 0 with size 3
---------------------------------------------------------------------------
```

However, you can change the value of an existing element. The following statement would work fine:

CODE:

```
a[0]=2
```

In summary, while you can modify the values of existing items in an array, you cannot add new items to it.

Now that we have seen how to define and reshape an array, we look at the ways in which we can combine arrays.

Combining arrays

There are three methods for combining arrays: appending, concatenation, and stacking.

1. **Appending** involves joining one array at the end of another array. The *np.append* function is used to append two arrays.

   CODE:

   ```
   x=np.array([[1,2],[3,4]])
   y=np.array([[6,7,8],[9,10,11]])
   np.append(x,y)
   ```

   Output:

   ```
   array([ 1,  2,  3,  4,  6,  7,  8,  9, 10, 11])
   ```

2. **Concatenation** involves joining arrays along an axis (either vertical or horizontal). The *np.concatenate* function concatenates arrays.

   CODE:

   ```
   x=np.array([[1,2],[3,4]])
   y=np.array([[6,7],[9,10]])
   np.concatenate((x,y))
   ```

Output:

```
array([[ 1,  2],
       [ 3,  4],
       [ 6,  7],
       [ 9, 10]])
```

By default, the *concatenate* function joins the arrays vertically (along the "0" axis). If you want the arrays to be joined side by side, the "axis" parameter needs to be added with the value as "1":

CODE:

```
np.concatenate((x,y),axis=1)
```

The *append* function uses the *concatenate* function internally.

3. **Stacking**: Stacking can be of two types, vertical or horizontal, as explained in the following.

   Vertical stacking

   As the name indicates, vertical stacking stacks arrays one below the other. The number of elements in each subarray of the arrays being stacked vertically must be the same for vertical stacking to work. The *np.vstack* function is used for vertical stacking.

   CODE:

```
x=np.array([[1,2],[3,4]])
y=np.array([[6,7],[8,9],[10,11]])
np.vstack((x,y))
```

   Output:

```
array([[ 1,  2],
       [ 3,  4],
       [ 6,  7],
       [ 8,  9],
       [10, 11]])
```

See how there are two elements in each subarray of the arrays "x" and "y".

Horizontal stacking

Horizontal stacking stacks arrays side by side. The number of subarrays needs to be the same for each of the arrays being horizontally stacked. The *np.hstack* function is used for horizontal stacking.

In the following example, we have two subarrays in each of the arrays, "x" and "y".

CODE:

```
x=np.array([[1,2],[3,4]])
y=np.array([[6,7,8],[9,10,11]])
np.hstack((x,y))
```

Output:

```
array([[ 1,  2,  6,  7,  8],
[ 3,  4,  9, 10, 11]])
```

In the next section, we look at how to use logical operators to test for conditions in NumPy arrays.

Testing for conditions

NumPy uses logical operators (&,|,~), and functions like *np.any, np.all,* and *np.where* to check for conditions. The elements in the array (or their indexes) that satisfy the condition are returned.

Consider the following array:

CODE:

```
x=np.linspace(1,50,10)
x
```

Output:

```
array([ 1.        ,  6.44444444, 11.88888889, 17.33333333, 22.77777778,
       28.22222222, 33.66666667, 39.11111111, 44.55555556, 50.        ])
```

Let us check for the following conditions and see which elements satisfy them:

- Checking if all the values satisfy a given condition: The *np.all* function returns the value "True" only if the condition holds for all the items of the array, as shown in the following example.

  CODE:

  ```
  np.all(x>20)
  #returns True only if all the elements are greater than 20
  ```

  Output:

  ```
  False
  ```

- Checking if any of the values in the array satisfy a condition: The *np.any* function returns the value "True" if any of the items satisfy the condition.

  CODE:

  ```
  np.any(x>20)
  #returns True if any one element in the array is greater than 20
  ```

  Output:

  ```
  True
  ```

- Returning the index of the items satisfy a condition: The *np.where* function returns the index of the values in the array satisfying a given condition.

  CODE:

  ```
  np.where(x<10)
  #returns the index of elements that are less than 10
  ```

Output:

```
(array([0, 1], dtype=int64),)
```

The *np.where* function is also useful for selectively retrieving or filtering values in an array. For example, we can retrieve those items that satisfy the condition "x<10", using the following code statement:

CODE:

```
x[np.where(x<10)]
```

Output:

```
array([1.        , 6.44444444])
```

- Checking for more than one condition:

  NumPy uses the following Boolean operators to combine conditions:

  - & operator (equivalent to *and* operator in Python): Returns True when all conditions are satisfied:

    CODE:

    ```
    x[(x>10) & (x<50)]
    #Returns all items that have a value greater than 10 and less
    than 50
    ```

    Output:

    ```
    array([11.88888889, 17.33333333, 22.77777778, 28.22222222,
    33.66666667,
    39.11111111, 44.55555556])
    ```

  - | operator (equivalent to *or* operator in Python): Returns True when any one condition, from a given set of conditions, is satisfied.

    CODE:

    ```
    x[(x>10) | (x<5)]
    #Returns all items that have a value greater than 10 or less
    than 5
    ```

Output:

```
array([ 1.        , 11.88888889, 17.33333333, 22.77777778,
28.22222222,
       33.66666667, 39.11111111, 44.55555556, 50.        ])
```

- ~ operator (equivalent to *not* operator in Python) for negating a condition.

  CODE:

```
x[~(x<8)]
#Returns all items greater than 8
```

  Output:

```
array([11.88888889, 17.33333333, 22.77777778, 28.22222222,
33.66666667,
39.11111111, 44.55555556, 50.        ])
```

We now move on to some other important concepts in NumPy like broadcasting and vectorization. We also discuss the use of arithmetic operators with NumPy arrays.

Broadcasting, vectorization, and arithmetic operations

Broadcasting

When we say that two arrays can be broadcast together, this means that their dimensions are compatible for performing arithmetic operations on them. Arrays can be combined using arithmetic operators as long as the rules of broadcasting are followed, which are explained in the following.

1. Both the arrays have the same dimensions.

   In this example, both arrays have the dimensions 2*6.

   CODE:

```
x=np.arange(0,12).reshape(2,6)
y=np.arange(5,17).reshape(2,6)
x*y
```

Output:

```
array([[  0,   6,  14,  24,  36,  50],
[ 66,  84, 104, 126, 150, 176]])
```

2. One of the arrays is a one-element array.

   In this example, the second array has only one element.

   CODE:

```
x=np.arange(0,12).reshape(2,6)
y=np.array([1])
x-y
```

   Output:

```
array([[-1,  0,  1,  2,  3,  4],
[ 5,  6,  7,  8,  9, 10]])
```

3. An array and a scalar (a single value) are combined.

   In this example, the variable y is used as a scalar value in the operation.

   CODE:

```
x=np.arange(0,12).reshape(2,6)
y=2
x/y
```

   Output:

```
array([[0. , 0.5, 1. , 1.5, 2. , 2.5],
[3. , 3.5, 4. , 4.5, 5. , 5.5]])
```

We can add, subtract, multiply, and divide arrays using either the arithmetic operators (+/-/* and /), or the functions (*np.add, np.subtract, np.multiply,* and *np.divide*)

CODE:

```
np.add(x,y)
#Or
x+y
```

Output:

```
array([[ 6,  8],
       [11, 13]])
```

Similarly, you can use *np.subtract* (or the – operator) for subtraction, *np.multiply* (or the * operator) for multiplication, and *np.divide* (or the / operator) for division.

Further reading: See more on array broadcasting: `https://numpy.org/doc/stable/user/basics.broadcasting.html`

Vectorization

Using the principle of vectorization, you can also conveniently apply arithmetic operators on each object in the array, instead of iterating through the elements, which is what you would do for applying operations to items in a container like a list.

CODE:

```
x=np.array([2,4,6,8])
x/2
#divides each element by 2
```

Output:

```
array([1., 2., 3., 4.])
```

Dot product

We can obtain the dot product of two arrays, which is different from multiplying two arrays. Multiplying two arrays gives an element-wise product, while a dot product of two arrays computes the inner product of the elements.

If we take two arrays,

```
|PQ|
|RS|
and
|UV|
|WX|
```

The dot product is given by

```
|PQ| . |UV| = |P*U+Q*VP*V+Q*X|
|R S|   |WX|    |R*U+S*WR*V+S*X|
```

Multiplying the arrays gives the following result:

```
|PQ| * |UV| = |P*U Q*V|
|R S|   |WX|    |R*WS*X|
```

As discussed earlier, arrays can be multiplied using the multiplication operator (*) or the *np.multiply* function.

The NumPy function for obtaining the dot product is *np.dot.*

CODE:

```
np.dot(x,y)
```

Output:

```
array([[21, 24],
       [47, 54]])
```

We can also combine an array with a scalar.

In the next topic, we discuss how to obtain the various properties or attributes of an array.

Obtaining the properties of an array

Array properties like their size, dimensions, number of elements, and memory usage can be found out using attributes.

Consider the following array:

```
x=np.arange(0,10).reshape(5,2)
```

- The *size* property gives the number of elements in the array.

  CODE:

  ```
  x.size
  ```

Output:

```
10
```

- The *ndim* property gives the number of dimensions.

  CODE:

  ```
  x.ndim
  ```

  Output:

  ```
  2
  ```

- The memory (total number of bytes) occupied by an array can be calculated using the *nbytes* attribute.

  CODE:

  ```
  x.nbytes
  ```

  Output:

  ```
  40
  ```

  Each element occupies 4 bytes (since this is an int array); therefore, ten elements occupy 40 bytes

- The data type of elements in this array can be calculated using the *dtype* attribute.

  CODE:

  ```
  x.dtype
  ```

  Output:

  ```
  dtype('int32')
  ```

Note the difference between the *dtype* and the *type* of an array. The *type* function gives the type of the container object (in this case, the type is *ndarray*), and dtype, which is an attribute, gives the type of individual items in the array.

Further reading: Learn more about the list of data types supported by NumPy: `https://numpy.org/devdocs/user/basics.types.html`

Transposing an array

The transpose of an array is its mirror image.

Consider the following array:

CODE:

```
x=np.arange(0,10).reshape(5,2)
```

There are two methods for transposing an array:

- We can use the *np.transpose* method.

  CODE:

  ```
  np.transpose(x)
  ```

- Alternatively, we can use the T attribute to obtain the transpose.

  CODE:

  ```
  x.T
  ```

  Both methods give the same output:

  ```
  array([[0, 2, 4, 6, 8],
  [1, 3, 5, 7, 9]])
  ```

Masked arrays

Let us say that you are using a NumPy array to store the scores obtained in an exam for a class of students. While you have data for most students, there are some missing values. A masked array, which is used for storing data with invalid or missing entries, is useful in such scenarios.

A masked array can be defined by creating an object of the "ma.masked_array" class (part of the numpy.ma module):

CODE:

```
import numpy.ma as ma
x=ma.masked_array([87,99,100,76,0],[0,0,0,0,1])
#The last element is invalid or masked
x[4]
```

Output:

```
Masked
```

Two arrays are passed as arguments to the *ma.masked_array* class – one containing the values of the items in the array, and one containing the mask values. A mask value of "0" indicates that the corresponding item value is valid, and a mask value of "1" indicates that it is missing or invalid. For instance, in the preceding example, the values 87, 99, 100, and 76 are valid since they have a mask value of "0". The last item in the first array (0), with a mask value of "1", is invalid.

The mask values can also be defined using the mask attribute.

CODE:

```
x=ma.array([87,99,100,76,0])
x.mask=[0,0,0,0,1]
```

To unmask an element, assign it a value:

CODE:

```
x[4]=82
```

The mask value for this element changes to 1 since it is no longer invalid.

Let us now look at how to create subsets from an array.

Slicing or selecting a subset of data

Slicing of arrays is similar to the slicing of strings and lists in Python. A slice is a subset of a data structure (in this case, an array), which can represent a set of values or a single value.

Consider the following array:

CODE:

```
x=np.arange(0,10).reshape(5,2)
```

Output:

```
array([[0, 1],
       [2, 3],
       [4, 5],
       [6, 7],
       [8, 9]])
```

Some examples of slicing are given in the following.

- Select the first subarray [0,1]:

  CODE:

  ```
  x[0]
  ```

  Output:

  ```
  array([0, 1])
  ```

- Select the second column:

  CODE:

  ```
  x[:,1]
  #This will select all the rows and the 2ndcolumn (has an
  index of 1)
  ```

  Output:

  ```
  array([1, 3, 5, 7, 9])
  ```

- Select the element at the fourth row and first column:

  CODE:

  ```
  x[3,0]
  ```

  Output:

  ```
  6
  ```

- We can also create a slice based on a condition:

  CODE:

  ```
  x[x<5]
  ```

Output:

```
array([0, 1, 2, 3, 4])
```

When we slice an array, the original array is not modified (a copy of the array is created).

Now that we have learned about creating and working with arrays, we move on to another important application of NumPy - calculation of statistical measures using various functions.

Obtaining descriptive statistics/aggregate measures

There are methods in NumPy that enable simplification of complex calculations and determination of aggregate measures.

Let us find the measures of central tendency (the mean, variance, standard deviation), sum, cumulative sum, and the maximum value for this array:

CODE:

```
x=np.arange(0,10).reshape(5,2)
#mean
x.mean()
```

Output:

```
4.5
```

Finding out the variance:

CODE:

```
x.var() #variance
```

Output:

```
2.9166666666666665
```

Calculating the standard deviation:

CODE:

```
x.std()  #standard deviation
```

Output:

```
1.707825127659933
```

Calculating the sum for each column:

CODE:

```
x.sum(axis=0) #calculates the column-wise sum
```

Output:

```
array([ 6, 15])
```

Calculating the cumulative sum:

CODE:

```
x.cumsum()
#calculates the sum of 2 elements at a time and adds this sum to the next
element
```

Output:

```
array([ 0,  1,  3,  6, 10, 15, 21, 28, 36, 45], dtype=int32)
```

Finding out the maximum value in an array:

CODE:

```
x.max()
```

Output:

```
9
```

Before concluding the chapter, let us learn about matrices – another data structure supported by the NumPy package.

Matrices

A matrix is a two-dimensional data structure, while an array can consist of any number of dimensions.

With the *np.matrix* class, we can create a matrix object, using the following syntax:

CODE:

```
x=np.matrix([[2,3],[33,3],[4,1]])
#OR
x=np.matrix('2,3;33,3;4,1') #Using semicolons to separate the rows
x
```

Output:

```
matrix([[ 2,  3],
        [33,  3],
        [ 4,  1]])
```

Most of the functions that can be applied to arrays can be used on matrices as well. Matrices use some arithmetic operators that make matrix operations more intuitive. For instance, we can use the * operator to get the dot product of two matrices that replicates the functionality of the *np.dot* function.

Since matrices are just one specific case of arrays and might be deprecated in future releases of NumPy, it is generally preferable to use NumPy arrays.

Summary

- NumPy is a library used for mathematical computations and creating data structures, called arrays, that can contain any number of dimensions.
- There are multiple ways for creating an array, and arrays can also be reshaped to add more dimensions or change the existing dimensions.

- Arrays support vectorization that provides a quick and intuitive method to apply arithmetic operators on all the elements of the array.
- A variety of statistical and aggregate measures can be calculated using simple NumPy functions, like *np.mean, np.var, np.std,* and so on.

Review Exercises

Question 1

Create the following array:

```
array([[[ 1,  2,  3,  4],
        [ 5,  6,  7,  8]],

       [[ 9, 10, 11, 12],
        [13, 14, 15, 16]],

       [[17, 18, 19, 20],
        [21, 22, 23, 24]]])
```

Slice the preceding array to obtain the following:

- Elements in the third subarray (17,18,19,20,21,22,23,24)
- Last element (24)
- Elements in the second column (2,6,10,14,18,22)
- Elements along the diagonal (1,10,19,24)

Question 2

Use the appropriate NumPy function for creating each of these arrays:

- An array with seven random numbers
- An uninitialized 2*5 array
- An array with ten equally spaced floating-point numbers between 1 and 3
- A 3*3 array with all the values as 100

Question 3

Write simple code statements for each of the following:

- Create an array to store the first 50 even numbers
- Calculate the mean and standard deviation of these numbers
- Reshape this array into an array with two subarrays, each with five rows and five columns
- Calculate the dimensions of this reshaped array

Question 4

Compute the dot product of these two data structures:

```
[[ 2,  3],
[33,  3],
[ 4,  1]]
```

AND

```
[[ 2,  3, 33],
[ 3,  4,  1]]
```

Using

1. Matrices
2. Arrays

Question 5

What is the difference between the code written in parts 1 and 2, and how would the outputs differ?

Part 1:

CODE:

```
x=np.array([1,2,3])
x*3
```

Part 2:

CODE:

```
a=[1,2,3]
a*3
```

Answers

Question 1

```
x=np.arange(1,25).reshape(3,2,4)
```

- Elements in the third subarray (17,18,19,20,21,22,23,24):

  CODE:

  ```
  x[2]
  ```

- Last element (24):

  CODE:

  ```
  x[2,1,3]
  ```

- Elements in the second column (2,6,10,14,18,22):

  CODE:

  ```
  x[:,:,1]
  ```

- Elements along the diagonal (1,10,19,24):

  CODE:

  ```
  x[0,0,0],x[1,0,1],x[2,0,2],x[2,1,3]
  ```

Question 2

- An array with seven random numbers:

  CODE:

  ```
  np.random.randn(7)
  ```

- An uninitialized 2*5 array:

  CODE:

  ```
  np.empty((2,5))
  ```

- An array with ten equally spaced floating-point numbers between 1 and 3:

  CODE:

  ```
  np.linspace(1,3,10)
  ```

- A 3*3 array with all the values as 100:

  CODE:

  ```
  np.full((3,3),100)
  ```

Question 3

CODE:

```
#creating the array of first 50 even numbers
x=np.arange(2,101,2)
#calculating the mean
x.mean()
#calculating the standard deviation
x.std()
#reshaping the array
y=x.reshape(2,5,5)
#calculating its new dimensions
y.ndim
```

Question 4

Computing the dot product using matrices requires the use of the * arithmetic operator

CODE:

```
x=np.matrix([[2,3],[33,3],[4,1]])
y=np.matrix([[2,3,33],[3,4,1]])
x*y
```

Computing the dot product using arrays requires the use of the dot method.

CODE:

```
x=np.array([[2,3],[33,3],[4,1]])
y=np.array([[2,3,33],[3,4,1]])
x.dot(y)
```

Question 5

Outputs:

1. `array([3, 6, 9])`

   An array supports vectorization, and thus the * operator is applied to each element.

2. `[1, 2, 3, 1, 2, 3, 1, 2, 3]`

   For a list, vectorization is not supported, and applying the * operator simply repeats the list instead of multiplying the elements by a given number. A "for" loop is required to apply an arithmetic operator on each item.

CHAPTER 6

Prepping Your Data with Pandas

With the explosion of the Internet, social networks, mobile devices, and big data, the amount of data available is humongous. Managing and analyzing this data to derive meaningful inferences can drive decision making, improve productivity, and reduce costs. In the previous chapter, you learned about NumPy – the library that helps us work with arrays and perform computations, also serving as the backbone for the Pandas library that we discuss in this chapter. Pandas, the Python library for data wrangling, has the advantage of being a powerful tool with many capabilities to manipulate data.

The rising popularity of Python as the preferred programming language is closely related to its widespread applications in the field of data science. In a survey conducted in 2019 among Python developers, it was found that NumPy and Pandas are the most popular data science frameworks (Source: `https://www.jetbrains.com/lp/python-developers-survey-2019/`).

In this chapter, we learn about the building blocks of the Pandas (Series, DataFrames, and Indexes), and understand the various functions in this library that are used to tidy, cleanse, merge, and aggregate data in Pandas. This chapter is more involved than the other chapters you have read so far since we are covering a wide range of topics that will help you develop the skills necessary for preparing your data.

Pandas at a glance

Wes McKinney developed the Pandas library in 2008. The name (Pandas) comes from the term "Panel Data" used in econometrics for analyzing time-series data. Pandas has many features, listed in the following, that make it a popular tool for data wrangling and analysis.

G. Rajagopalan, *A Python Data Analyst's Toolkit*, https://doi.org/10.1007/978-1-4842-6399-0_6

1. Pandas provides features for labeling of data or indexing, which speeds up the retrieval of data.

2. Input and output support: Pandas provides options to read data from different file formats like JSON (JavaScript Object Notation), CSV (Comma-Separated Values), Excel, and HDF5 (Hierarchical Data Format Version 5). It can also be used to write data into databases, web services, and so on.

3. Most of the data that is needed for analysis is not contained in a single source, and we often need to combine datasets to consolidate the data that we need for analysis. Again, Pandas comes to the rescue with tailor-made functions to combine data.

4. Speed and enhanced performance: The Pandas library is based on Cython, which combines the convenience and ease of use of Python with the speed of the C language. Cython helps to optimize performance and reduce overheads.

5. Data visualization: To derive insights from the data and make it presentable to the audience, viewing data using visual means is crucial, and Pandas provides a lot of built-in visualization tools using Matplotlib as the base library.

6. Support for other libraries: Pandas integrates smoothly with other libraries like Numpy, Matplotlib, Scipy, and Scikit-learn. Thus we can perform other tasks like numerical computations, visualizations, statistical analysis, and machine learning in conjunction with data manipulation.

7. Grouping: Pandas provides support for the split-apply-combine methodology, whereby we can group our data into categories, apply separate functions on them, and combine the results.

8. Handling missing data, duplicates, and filler characters: Data often has missing values, duplicates, blank spaces, special characters (like $, &), and so on that may need to be removed or replaced. With the functions provided in Pandas, you can handle such anomalies with ease.

9. Mathematical operations: Many numerical operations and computations can be performed in Pandas, with NumPy being used at the back end for this purpose.

Technical requirements

The libraries and the external files needed for this chapter are detailed in the following.

Installing libraries

If you have not already installed Pandas, go to the Anaconda Prompt and enter the following command.

```
>>>pip install pandas
```

Once the Pandas library is installed, you need to import it before using its functions. In your Jupyter notebook, type the following to import this library.

CODE:

```
import pandas as pd
```

Here, *pd* is a shorthand name or alias that is a standard for Pandas.

For some of the examples, we also use functions from the NumPy library. Ensure that both the Pandas and NumPy libraries are installed and imported.

External files

You need to download a dataset, "`subset-covid-data.csv`", that contains data about the number of cases and deaths related to the COVID-19 pandemic for various countries on a particular date. Please use the following link for downloading the dataset: `https://github.com/DataRepo2019/Data-files/blob/master/subset-covid-data.csv`

Building blocks of Pandas

The Series and DataFrame objects are the underlying data structures in Pandas. In a nutshell, a Series is like a column (has only one dimension), and a DataFrame (has two dimensions) is like a table or a spreadsheet with rows and columns. Each value stored in a Series or a DataFrame has a label or an index attached to it, which speeds up retrieval and access to data. In this section, we learn how to create a Series and DataFrame, and the functions used for manipulating these objects.

Creating a Series object

The Series is a one-dimensional object, with a set of values and their associated indexes. Table 6-1 lists the different ways of creating a series.

Table 6-1. *Various Methods for Creating a Series Object*

METHOD	SYNTAX

Using a scalar value	CODE (for creating a Series using a scalar value):
	`pd.Series(2)`
	`#Creating a simple series with just one value. Here, 0 is the index label, and 2 is the value the Series object contains.`
	Output:
	`0    2`
	`dtype: int64`

Using a list	CODE (for creating a Series using a list):
	`pd.Series([2]*5)`
	`#Creating a series by enclosing a single value (2) in a list and replicating it 5 times. 0,1,2,3,4 are the autogenerated index labels.`
	Output:
	`0    2`
	`1    2`
	`2    2`
	`3    2`
	`4    2`
	`dtype: int64`

Using characters in a string	CODE (for creating a series using a string):
	`pd.Series(list('hello'))`
	`#Creating a series by using each character in the string "hello" as a separate value in the Series.`

(*continued*)

Table 6-1. (*continued*)

METHOD	SYNTAX

	Output:
	`0    h`
	`1    e`
	`2    l`
	`3    l`
	`4    o`
	`dtype: object`

Using a dictionary	CODE (for creating a Series from a dictionary):
	`pd.Series({1:'India',2:'Japan',3:'Singapore'})`
	#The key/value pairs correspond to the index labels and values in the Series object.
	
	Output:
	`1        India`
	`2        Japan`
	`3    Singapore`
	`dtype: object`

Using a range	CODE (for creating a Series from a range):
	`pd.Series(np.arange(1,5))`
	`#Using the NumPy arrange function to create a series from a range of 4 numbers (1-4), ensure that the NumPy library is also imported`
	
	Output:
	`0    1`
	`1    2`
	`2    3`
	`3    4`
	`dtype: int32`

(*continued*)

Table 6-1. (*continued*)

METHOD	SYNTAX

Using random numbers	CODE (for creating a Series from random numbers):
	`pd.Series(np.random.normal(size=4))`
	`#Creating a set of 4 random numbers using the np.random.normal function`
	
	Output:
	`0    -1.356631`
	`1     1.308935`
	`2    -1.247753`
	`3    -1.408781`
	`dtype: float64`

Creating a series with customized index labels	CODE (for creating a custom index):
	`pd.Series([2,0,1,6],index=['a','b','c','d'])`
	`#The list [2,0,1,6] specifies the values in the series, and the list for the index['a','b','c','d'] specifies the index labels`
	
	Output:
	`a    2`
	`b    0`
	`c    1`
	`d    6`
	`dtype: int64`

To summarize, you can create a Series object from a single (scalar) value, list, dictionary, a set of random numbers, or a range of numbers. The *pd.Series* function creates a Series object (note that the letter "S" in "Series" is in uppercase; pd.series will not work). Use the index parameter if you want to customize the index.

Examining the properties of a Series

In this section, we will look at the methods used for finding out more information about a Series object like the number of elements, its values, and unique elements.

Finding out the number of elements in a Series

There are three ways of finding the number of elements a Series contains: using the *size* parameter, the *len* function, or the *shape* parameter

The *size* attribute and the *len* function return a single value - the length of the series, as shown in the following.

CODE:

```
#series definition
x=pd.Series(np.arange(1,10))
#using the size attribute
x.size
```

Output:

```
9
```

We can also use the *len* function for calculating the number of elements, which would return the same output (9), as shown in the following.

CODE:

```
len(x)
```

The *shape* attribute returns a tuple with the number of rows and columns. Since the Series object is one-dimensional, the shape attribute returns only the number of rows, as shown in the following.

CODE:

```
x.shape
```

Output:

```
(9,)
```

Listing the values of the individual elements in a Series

The *values* attribute returns a NumPy array containing the values of each item in the Series.

CODE:

```
x.values
```

Output:

```
array([1, 2, 3, 4, 5, 6, 7, 8, 9])
```

Accessing the index of a Series

The index of the Series can be accessed through the index attribute. An index is an object with a data type and a set of values. The default type for an index object is *RangeIndex*.

CODE:

```
x.index
```

Output:

```
RangeIndex(start=0, stop=9, step=1)
```

The index labels form a range of numbers, starting from 0. The default step size or the difference between one index label value and the next is 1.

Obtaining the unique values in a Series and their count

The *value_counts()* method is an important method. When used with a Series object, it displays the unique values contained in this object and the count of each of these unique values. It is a common practice to use this method with a categorical variable, to get an idea of the distinct values it contains.

CODE:

```
z=pd.Series(['a','b','a','c','d','b'])
z.value_counts()
```

Output:

```
a    2
b    2
c    1
d    1
dtype: int64
```

The preceding output shows that in the Series object named "z", the values "a" and "b" occur twice, while the characters "c" and "d" occur once.

Method chaining for a Series

We can apply multiple methods to a series and apply them successively. This is called method chaining and can be applied for both Series and DataFrame objects.

Example:

Suppose we want to find out the number of times the values "a" and "b" occur for the series "z" defined in the following. We can combine the *value_counts* method and the *head* method by chaining them.

CODE:

```
z=pd.Series(['a','b','a','c','d','b'])
z.value_counts().head(2)
```

Output:

```
a    2
b    2
dtype: int64
```

If multiple methods need to be changed together and applied on a Series object, it is better to mention each method on a separate line, with each line ending with a backslash. It would make the code more readable, as shown in the following.

CODE:

```
z.value_counts()\
.head(2)\
.values
```

Output:

```
array([2, 2], dtype=int64)
```

We have covered the essential methods that are used with the Series object. If you want to learn more about the Series object and the methods used with Series objects, refer to the following link.

https://pandas.pydata.org/pandas-docs/stable/reference/api/pandas.Series.html

We now move ovn to DataFrames, another important Pandas object.

DataFrames

A DataFrame is an extension of a Series. It is a two-dimensional data structure for storing data. While the Series object contains two components - a set of values, and index labels attached to these values - the DataFrame object contains three components - the column object, index object, and a NumPy array object that contains the values.

The index and columns are collectively called the axes. The index forms the axis "0" and the columns form the axis "1".

We look at various methods for creating DataFrames in Table 6-2.

Table 6-2. *Different Methods for Creating a DataFrame*

<table>
<tr><th>Method</th><th>Syntax</th></tr>
<tr><td>By combining Series objects</td><td>CODE:
<pre>student_ages=pd.Series([22,24,20]) #series 1
teacher_ages=pd.Series([40,50,45])#series 2
combined_ages=pd.DataFrame([student_ages,
teacher_ages]) #DataFrame
combined_ages.columns=['class 1','class 2',
'class 3']#naming columns
combined_ages</pre>
Output:
<table>
<tr><th></th><th>class 1</th><th>class 2</th><th>class 3</th></tr>
<tr><td>0</td><td>22</td><td>24</td><td>20</td></tr>
<tr><td>1</td><td>40</td><td>50</td><td>45</td></tr>
</table>
Here, we are defining two Series and then using the pd.DataFrame function to create a new DataFrame called "combined_ages". We give names to columns in a separate step.</td></tr>
</table>

(continued)

Table 6-2. (*continued*)

<table>
<tr><th>Method</th><th>Syntax</th></tr>
<tr><td>From a dictionary</td><td>CODE:
<pre>combined_ages=pd.DataFrame({'class 1':[22,40],'class 2':[24,50],'class 3':[20,45]})
combined_ages</pre>Output:
<table><tr><th></th><th>class 1</th><th>class 2</th><th>class 3</th></tr><tr><td>0</td><td>22</td><td>24</td><td>20</td></tr><tr><td>1</td><td>40</td><td>50</td><td>45</td></tr></table>A dictionary is passed as an argument to the pd.DataFrame function (with the column names forming keys, and values in each column enclosed in a list).</td></tr>
<tr><td>From a numpy array</td><td>CODE:
<pre>numerical_df=pd.DataFrame(np.arange(1,9).reshape(2,4))
numerical_df</pre>Output:
<table><tr><th></th><th>0</th><th>1</th><th>2</th><th>3</th></tr><tr><td>0</td><td>1</td><td>2</td><td>3</td><td>4</td></tr><tr><td>1</td><td>5</td><td>6</td><td>7</td><td>8</td></tr></table>Here, we create a NumPy array first using the np.arange function. Then we reshape this array into a DataFrame with two rows and four columns.</td></tr>
</table>

(*continued*)

Table 6-2. *(continued)*

<table>
<tr><th>Method</th><th>Syntax</th></tr>
<tr><td>Using a set of tuples</td><td>CODE:
combined_ages=pd.DataFrame([(22,24,20),(40,50,45)],col
umns=['class 1','class 2','class 3'])
combined_ages
Output:
<table>
<tr><th></th><th>class 1</th><th>class 2</th><th>class 3</th></tr>
<tr><td>0</td><td>22</td><td>24</td><td>20</td></tr>
<tr><td>1</td><td>40</td><td>50</td><td>45</td></tr>
</table>
We have re-created the "combined_ages" DataFrame using a set of tuples. Each tuple is equivalent to a row in a DataFrame.</td></tr>
</table>

To sum up, we can create a DataFrame using a dictionary, a set of tuples, and by combining Series objects. Each of these methods uses the *pd.DataFrame* function. Note that the characters "D" and "F" in this method are in uppercase; pd.dataframe does not work.

Creating DataFrames by importing data from other formats

Pandas can read data from a wide variety of formats using its reader functions (refer to the complete list of supported formats here: `https://pandas.pydata.org/pandas-docs/stable/user_guide/io.html`). The following are some of the commonly used formats.

From a CSV file:

The *read_csv* function can be used to read data from a CSV file into a DataFrame, as shown in the following.

CODE:

```
titanic=pd.read_csv('titanic.csv')
```

Reading data from CSV files is one of the most common ways to create a DataFrame. CSV files are comma-separated files for storing and retrieving values, where each line is equivalent to a row. Remember to upload the CSV file in Jupyter using the upload button on the Jupyter home page (Figure 6-1), before calling the "read_csv" function.

Figure 6-1. Jupyter file upload

From an Excel file:

Pandas provides support for importing data from both xls and xlsx file formats using the *pd.read_excel* function, as shown in the following.

CODE:

```
titanic_excel=pd.read_excel('titanic.xls')
```

From a JSON file:

JSON stands for JavaScript Object Notation and is a cross-platform file format for transmitting and exchanging data between the client and server. Pandas provides the function *read_json* to read data from a JSON file, as shown in the following.

CODE:

```
titanic=pd.read_json('titanic-json.json')
```

From an HTML file:

We can also import data from a web page using the *pd.read_html* function.

In the following example, this function parses the tables on the web page into DataFrame objects. This function returns a list of DataFrame objects which correspond to the tables on the web page. In the following example, table[0] corresponds to the first table on the mentioned URL.

CODE:

```
url="https://www.w3schools.com/sql/sql_create_table.asp"
table=pd.read_html(url)
table[0]
```

Output:

	PersonID	LastName	FirstName	Address	City
0	NaN	NaN	NaN	NaN	NaN

Further reading: See the complete list of supported formats in Pandas and the functions for reading data from such formats:

https://pandas.pydata.org/pandas-docs/stable/reference/io.html

Accessing attributes in a DataFrame

In this section, we look at how to access the attributes in a DataFrame object.

We use the following DataFrame:

CODE:

```
combined_ages=pd.DataFrame({'class 1':[22,40],'class 2':[24,50],
'class 3':[20,45]})
```

Attributes

The index attribute, when used with a DataFrame object, gives the type of an index object and its values.

CODE:

```
combined_ages.index
```

Output:

```
RangeIndex(start=0, stop=2, step=1)
```

The columns attribute gives you information about the columns (their names and data type).

CODE:

```
combined_ages.columns
```

Output:

```
Index(['class 1', 'class 2', 'class 3'], dtype="object")
```

The index object and column object are both types of index objects. While the index object has a type *RangeIndex*, the columns object has a type "Index". The values of the index object act as row labels, while those of the column object act as column labels.

Accessing the values in the DataFrame

Using the values attribute, you can obtain the data stored in the DataFrame. The output, as you can see, is an array containing the values.

CODE:

```
combined_ages.values
```

Output:

```
array([[22, 24, 20],
       [40, 50, 45]], dtype=int64)
```

Modifying DataFrame objects

In this section, we will learn how to change the names of columns and add and delete columns and rows.

Renaming columns

The names of the columns can be changed using the *rename* method. A dictionary is passed as an argument to this method. The keys for this dictionary are the old column names, and the values are the new column names.

CODE:

```
combined_ages.rename(columns={'class 1':'batch 1','class 2':'batch
2','class 3':'batch 3'},inplace=True)
combined_ages
```

Output:

	batch 1	batch 2	batch 3
0	22	24	20
1	40	50	45

The reason we use the *inplace* parameter so that the changes are made in the actual DataFrame object.

Renaming can also be done by accessing the columns attribute directly and mentioning the new column names in an array, as shown in the following example.

CODE:

```
combined_ages.columns=['batch 1','batch 2','batch 3']
```

Renaming using the dictionary format is a more straightforward method for renaming columns, and the changes are made to the original DataFrame object. The disadvantage with this method is that one needs to remember the order of the columns in the DataFrame. When we used the *rename* method, we used a dictionary where we knew which column names we were changing.

Replacing values or observations in a DataFrame

The *replace* method can be used to replace values in a DataFrame. We can again use the dictionary format, with the key/value pair representing the old and new values. Here, we replace the value 22 with the value 33.

CODE:

```
combined_ages.replace({22:33})
```

Output:

	class 1	class 2	class 3
0	33	24	20
1	40	50	45

Adding a new column to a DataFrame

There are four ways to insert a new column in a DataFrame, as shown in Table 6-3.

Table 6-3. Adding a New Column to a DataFrame

<table>
<tr><th>Method of column insertion</th><th>Syntax</th></tr>
<tr><td>With the indexing operator, [ ]</td><td>CODE: <code>combined_ages['class 4']=[18,40]</code> <code>combined_ages</code> Output: |	class 1	class 2	class 3	class 4
0	22	24	20	18
1	40	50	45	40 | By mentioning the column name as a string within the indexing operator and assigning it values, we can add a column.</td></tr>
<tr><td>Using the insert method</td><td>CODE: <code>combined_ages.insert(2,'class 0',[18,35])</code> <code>combined_ages</code> Output: |	class 1	class 2	class 0	class 3
0	22	24	18	20
1	40	50	35	45 | The insert method can be used for adding a column. Three arguments need to be passed to this method, mentioned in the following. The first argument is the index where you want to insert the new column (in this case the index is 2, which means that the new column is added as the third column of our DataFrame) The second argument is the name of the new column you want to insert ("class 0" in this example) The third argument is the list containing the values of the new column (18 and 35 in this case) All the three parameters are mandatory for the insert method to be able to add a column successfully.</td></tr>
</table>

(*continued*)

Table 6-3. *(continued)*

<table>
<tr><th>Method of column insertion</th><th>Syntax</th></tr>
<tr><td>Using the loc indexer</td><td>CODE:
<pre>combined_ages.loc[:,'class 4']=[20,40]
combined_ages</pre>Output:
<table>
<tr><th></th><th>class 1</th><th>class 2</th><th>class 3</th><th>class 4</th></tr>
<tr><td>0</td><td>22</td><td>24</td><td>20</td><td>20</td></tr>
<tr><td>1</td><td>40</td><td>50</td><td>45</td><td>40</td></tr>
</table>
The loc indexer is generally used for retrieval of values in from Series and DataFrames, but it can also be used for inserting a column. In the preceding statement, all the rows are selected using the : operator. This operator is followed by the name of the column to be inserted. The values for this column are enclosed within a list.</td></tr>
<tr><td>Using the concat function</td><td>CODE:
<pre>class5=pd.Series([31,48])
combined_ages=pd.concat([combined_ages,class5],axis=1)
combined_ages</pre>Output:
<table>
<tr><th></th><th>class 1</th><th>class 2</th><th>class 3</th><th>0</th></tr>
<tr><td>0</td><td>22</td><td>24</td><td>20</td><td>31</td></tr>
<tr><td>1</td><td>40</td><td>50</td><td>45</td><td>48</td></tr>
</table>
First, the column to be added (“class5” in this case) is defined as a Series object. It is then added to the DataFrame object using the pd.concat function. The axis needs to be mentioned as “1” since the new data is being added along the column axis.</td></tr>
</table>

In summary, we can add a column to a DataFrame using the indexing operator, *loc* indexer, *insert* method, or *concat* function. The most straightforward and commonly used method for adding a column is by using the indexing operator [].

Inserting rows in a DataFrame

There are two methods for adding rows in a DataFrame, either by using the *append* method or with the *concat* function, as shown in Table 6-4.

Table 6-4. *Adding a New Row to a DataFrame*

Method for row insertion	Syntax

Using the *append* method	CODE:
	`combined_ages=combined_ages.append({'class 1':35,'class 2':33,'class 3':21},ignore_index=True)`
	`combined_ages`
	Output:
	<table><tr><th></th><th>class 1</th><th>class 2</th><th>class 3</th></tr><tr><td>0</td><td>22</td><td>24</td><td>20</td></tr><tr><td>1</td><td>40</td><td>50</td><td>45</td></tr><tr><td>2</td><td>35</td><td>33</td><td>21</td></tr></table>
	The argument to the *append* method- the data that needs to be added - is defined as a dictionary. This dictionary is then passed as an argument to the *append* method. Setting the *ignore_index=True* parameter prevents an error from being thrown. This parameter resets the index. While using the *append* method, we need to ensure that we either use the *ignore_index* parameter or give a name to a Series before appending it to a DataFrame. Note that the *append* method does not have an *inplace* parameter that would ensure that the changes reflect in the original object; hence we need to set the original object to point to the new object created using append, as shown in the preceding code.

(*continued*)

Table 6-4. (*continued*)

<table>
<tr><th>Method for row insertion</th><th>Syntax</th></tr>
<tr><td>Using the pd.concat function</td><td>CODE:
<pre>new_row=pd.DataFrame([{'class 1':32,'class 2':37,
'class 3':41}])
pd.concat([combined_ages,new_row])</pre>Output:
<table><tr><th></th><th>class 1</th><th>class 2</th><th>class 3</th></tr><tr><td>0</td><td>22</td><td>24</td><td>20</td></tr><tr><td>1</td><td>40</td><td>50</td><td>45</td></tr><tr><td>0</td><td>32</td><td>37</td><td>41</td></tr></table>
The pd.concat function is used to add new rows as shown in the preceding syntax. The new row to be added is defined as a DataFrame object. Then the pd.concat function is called and the names of the two DataFrames (the original DataFrame and the new row defined as a DataFrame) are passed as arguments.</td></tr>
</table>

In summary, we can use either the *append* method or *concat* function for adding rows to a DataFrame.

Deleting columns from a DataFrame

Three methods can be used to delete a column from a DataFrame, as shown in Table 6-5.

Table 6-5. Deleting a Column from a DataFrame

Method for deletion of column	Syntax

del function	CODE:
	`del combined_ages['class 3']`
	`combined_ages`
	Output:
	<table><tr><th></th><th>class 1</th><th>class 2</th></tr><tr><td>0</td><td>22</td><td>24</td></tr><tr><td>1</td><td>40</td><td>50</td></tr></table>
	The preceding statement deletes the last column (with the name,"class 3").
	Note that the deletion occurs inplace, that is, in the original DataFrame itself.

Using the *pop* method	CODE:
	`combined_ages.pop('class 2')`
	Output:
	`0    24`
	`1    50`
	`Name: class 2, dtype: int64`
	The *pop* method deletes a column inplace and returns the deleted column as a Series object

(*continued*)

Table 6-5. *(continued)*

Method for deletion of column	Syntax

Using the drop method	CODE:
	`combined_ages.drop(['class 1'],axis=1,inplace=True)`
	`combined_ages`
	Output:
	<table><tr><th></th><th>class 2</th><th>class 3</th></tr><tr><td>0</td><td>24</td><td>20</td></tr><tr><td>1</td><td>50</td><td>45</td></tr></table>
	The column(s) that needs to be dropped is mentioned as a string within a list, which is then passed as an argument to the *drop* method. Since the *drop* method removes rows (axis=0) by default, we need to specify the axis value as "1" if we want to remove a column.
	Unlike the *del* function and *pop* method, the deletion using the *drop* method does not occur in the original DataFrame object, and therefore, we need to add the *inplace* parameter.

To sum up, we can use the *del* function, *pop* method, or *drop* method to delete a column from a DataFrame.

Deleting a row from a DataFrame

There are two methods for removing rows from a DataFrame – either by using a Boolean selection or by using the drop method, as shown in Table 6-6.

Table 6-6. Deleting Row from a DataFrame

<table>
<tr><th>Method of row deletion</th><th>Syntax</th></tr>
<tr><td>Using a Boolean selection</td><td>CODE:
<code>combined_ages[~(combined_ages.values<50)]</code>
Output:
<table><tr><th></th><th>class 1</th><th>class 2</th><th>class 3</th></tr><tr><td>1</td><td>40</td><td>50</td><td>45</td></tr></table>We use the NOT operator (~) to remove the rows that we do not want. Here, we remove all values in the DataFrame that are less than 50.</td></tr>
<tr><td>Using the drop method</td><td>CODE:
<code>combined_ages.drop(1)</code>
Output:
<table><tr><th></th><th>class 1</th><th>class 2</th><th>class 3</th></tr><tr><td>0</td><td>22</td><td>24</td><td>20</td></tr></table>Here, we remove the second row, which has a row index of 1. If there is more than one row to be removed, we need to specify the indexes of the rows in a list.</td></tr>
</table>

Thus, we can use either a Boolean selection or the drop method to remove rows from a DataFrame. Since the drop method works with the removal of both rows and columns, it can be used uniformly. Remember to add the required parameters to the drop method. For removing columns, the *axis* (=1) parameter needs to be added. For changes to reflect in the original DataFrame, the *inplace* (=True) parameter needs to be included.

Indexing

Indexing is fundamental to Pandas and is what makes retrieval and access to data much faster compared to other tools. It is crucial to set an appropriate index to optimize performance. An index is implemented in NumPy as an immutable (cannot be modified) array and contains hashable objects. A hashable object is one that can be converted to an integer value based on its contents (similar to mapping in a dictionary). Objects with different values will have different hash values.

Pandas has two types of indexes - a row index (vertical) with labels attached to rows, and a column index with labels (column names) for every column.

Let us now explore index objects – their data types, their properties, and how they speed up access to data.

Type of an index object

An index object has a data type, some of which are listed here.

- Index: This is a generic index type; the column index has this type.
- RangeIndex: Default index type in Pandas (used when an index is not defined separately), implemented as a range of increasing integers. This index type helps with saving memory.
- Int64Index: An index type containing integers as labels. For this index type, the index labels need not be equally spaced, whereas this is required for an index of type RangeIndex.
- Float64Index: Contains floating-point numbers (numbers with a decimal point) as index labels.
- IntervalIndex: Contains intervals (for instance, the interval between two integers) as labels.
- CategoricalIndex: A limited and finite set of values.
- DateTimeIndex: Used to represent date and time, like in time-series data.
- PeriodIndex: Represents periods like quarters, months, or years.
- TimedeltaIndex: Represents duration between two periods of time or two dates.
- MultiIndex: Hierarchical index with multiple levels.

Further reading:

Learn more about types of indexes here: `https://pandas.pydata.org/pandas-docs/stable/reference/api/pandas.Index.html`

Creating a custom index and using columns as indexes

When a Pandas object is created, a default index is created of the type RangeIndex, as mentioned earlier. An index of this type has the first label value as 0 (which corresponds to the first item of the Pandas Series or DataFrame), and the second label as 1, following an arithmetic progression with a spacing of one integer.

We can set a customized index, using either the index parameter or attribute. In the Series and DataFrame objects we created earlier, we were just setting values for the individual items, and in the absence of labels for the index object, the default index (of type RangeIndex) was used.

We can use the index parameter when we define a Series or DataFrame to give custom values to the index labels.

CODE:

```
periodic_table=pd.DataFrame({'Element':['Hydrogen','Helium','Lithium',
'Beryllium','Boron']},index=['H','He','Li','Be','B'])
```

Output:

	Element
H	Hydrogen
He	Helium
Li	Lithium
Be	Beryllium
B	Boron

If we skip the index parameter during the creation of the object, we can set the labels using the index attribute, as shown here.

CODE:

```
periodic_table.index=['H','He','Li','Be','B']
```

The *set_index* method can be used to set an index using an existing column, as demonstrated in the following:

CODE:

```
periodic_table=pd.DataFrame({'Element':['Hydrogen','Helium','Lithium',
'Beryllium','Boron'],'Symbols':['H','He','Li','Be','B']})
periodic_table.set_index(['Symbols'])
```

Output:

	Element
Symbols	
H	Hydrogen
He	Helium
Li	Lithium
Be	Beryllium
B	Boron

The index can be made a column again or reset using the reset_index method:

CODE:

```
periodic_table.reset_index()
```

Output:

	index	Element	Symbols
0	0	Hydrogen	H
1	1	Helium	He
2	2	Lithium	Li
3	3	Beryllium	Be
4	4	Boron	B

We can also set the index when we read data from an external file into a DataFrame, using the *index_col* parameter, as shown in the following.

CODE:

```
titanic=pd.read_csv('titanic.csv',index_col='PassengerId')
titanic.head()
```

Output:

	Survived	Pclass	Name	Sex	Age	SibSp	Parch	Ticket	Fare	Cabin	Embarked
PassengerId											
1	0	3	Braund, Mr. Owen Harris	male	22.0	1	0	A/5 21171	7.2500	NaN	S
2	1	1	Cumings, Mrs. John Bradley (Florence Briggs Th...	female	38.0	1	0	PC 17599	71.2833	C85	C
3	1	3	Heikkinen, Miss. Laina	female	26.0	0	0	STON/O2. 3101282	7.9250	NaN	S
4	1	1	Futrelle, Mrs. Jacques Heath (Lily May Peel)	female	35.0	1	0	113803	53.1000	C123	S
5	0	3	Allen, Mr. William Henry	male	35.0	0	0	373450	8.0500	NaN	S

Indexes and speed of data retrieval

We know that indexes dramatically improve the speed of access to data. Let us understand this with the help of an example.

Consider the following DataFrame:

CODE:

```
periodic_table=pd.DataFrame({'Atomic Number':[1,2,3,4,5],'Element':
['Hydrogen','Helium','Lithium','Beryllium','Boron'],'Symbol':['H','He',
'Li','Be','B']})
```

Output:

	Atomic Number	Element	Symbol
0	1	Hydrogen	H
1	2	Helium	He
2	3	Lithium	Li
3	4	Beryllium	Be
4	5	Boron	B

Searching without using an index

Now, try retrieving the element with atomic number 2 without the use of an index and measure the time taken for retrieval using the *timeit* magic function. When the index is not used, a linear search is performed to retrieve an element, which is relatively time consuming.

CODE:

```
%timeit periodic_table[periodic_table['Atomic Number']==2]
```

Output:

```
1.66 ms ± 99.1 µs per loop (mean ± std. dev. of 7 runs, 1000 loops each)
```

Search using an index

Now, set the "Atomic Number" column as the index and use the *loc* indexer to see how much time the search takes now:

CODE:

```
new_periodic_table=periodic_table.set_index(['Atomic Number'])
%timeit new_periodic_table.loc[2]
```

Output:

```
281 µs ± 14.4 µs per loop (mean ± std. dev. of 7 runs, 1000 loops each)
```

The search operation, when performed without using an index, was of the order of milliseconds (around 1.66 ms). With the use of indexes, the time taken for the retrieval operation is now of the order of microseconds (281 µs), which is a significant improvement.

Immutability of an index

As mentioned earlier, the index object is immutable - once defined, the index object or its labels cannot be modified.

As an example, let us try changing one of the index labels in the periodic table DataFrame we just defined, as shown in the following. We get an error in the output since we are trying to operate on an immutable object.

CODE:

```
periodic_table.index[2]=0
```

Output:

```
---------------------------------------------------------------------------
TypeError                                 Traceback (most recent call last)
<ipython-input-24-cd2fece917cb> in <module>
----> 1periodic_table.index[2]=0

~\Anaconda3\lib\site-packages\pandas\core\indexes\base.py in __setitem__
(self, key, value)
   3936
   3937def __setitem__(self, key, value):
-> 3938raiseTypeError("Index does not support mutable operations")
   3939
   3940def __getitem__(self, key):

TypeError: Index does not support mutable operations
```

While the values of an Index object cannot be changed, we can retrieve information about the index using its attributes, like the values contained in the Index object, whether there are any null values, and so on.

Let us look at some of the index attributes with some examples:

Considering the column index in the following DataFrame:

CODE:

```
periodic_table=pd.DataFrame({'Element':['Hydrogen','Helium','Lithium','Bery
llium','Boron']},index=['H','He','Li','Be','B'])

column_index=periodic_table.columns
```

Some of the attributes of the column index are

1.values attribute: Returns the column names

CODE:

```
column_index.values
```

Output:

```
array(['Element'], dtype=object)
```

2.hasnans attribute: Returns a Boolean True or False value based on the presence of null values.

CODE:

```
column_index.hasnans
```

Output:

```
False
```

3.nbytes attribute: Returns the number of bytes occupied in memory

CODE:

```
column_index.nbytes
```

Output:

```
8
```

Further reading: For a complete list of attributes, refer to the following documentation: https://pandas.pydata.org/pandas-docs/stable/reference/api/pandas.Index.html

Alignment of indexes

When two Pandas objects are added, their index labels are checked for alignment. For items that have matching indexes, their values are added or concatenated. Where the indexes do not match, the value corresponding to that index in the resultant object is null (*np.NaN*).

Let us understand this with an example. Here, we see that the 0 index label in s1 does not have a match in s2, and the last index label (10) in s2 does not have a match in s1. These values equal null when the objects are combined. All other values, where the index labels align, are added together.

CODE:

```
s1=pd.Series(np.arange(10),index=np.arange(10))
s2=pd.Series(np.arange(10),index=np.arange(1,11))
s1+s2
```

Output:

```
0       NaN
1       1.0
2       3.0
3       5.0
4       7.0
5       9.0
6      11.0
7      13.0
8      15.0
9      17.0
10      NaN
dtype: float64
```

Set operations on indexes

We can perform set operations like union, difference, and symmetric difference on indexes from different objects.

Consider the following indexes, "i1" and "i2", created from two Series objects ("s1" and "s2") we created in the previous section:

CODE:

```
i1=s1.index
i2=s2.index
```

Union operation

All elements present in both sets are returned.

CODE:

```
i1.union(i2)
```

Output:

```
Int64Index([0, 1, 2, 3, 4, 5, 6, 7, 8, 9, 10], dtype="int64")
```

Difference operation

Elements present in one set, but not in the other, are returned.

CODE:

```
i1.difference(i2) #elements present in i1 but not in i2
```

Output:

```
Int64Index([0], dtype="int64")
```

Symmetric difference operation

Elements not common to the two sets are returned. This operation differs from the Difference operation in that it takes into the uncommon elements in both sets:

CODE:

```
i1.symmetric_difference(i2)
```

Output:

```
Int64Index([0, 10], dtype="int64")
```

You can also perform arithmetic operations on two index objects, as shown in the following.

CODE:

```
i1-i2
```

Output:

```
Int64Index([-1, -1, -1, -1, -1, -1, -1, -1, -1, -1], dtype="int64")
```

Data types in Pandas

The data types used in Pandas are derived from NumPy, except for the "category" data type for qualitative data, which is defined in Pandas. The common data types include

- object (for storing mixed data like numbers, strings, etc.)
- int64 (for integer values)
- float64 (for numbers with decimal points)

- Datetime (for storing date and time data)
- Category (for variables containing only a few distinct values, like 'True'/'False', or some limited ordered categories like 'one'/'two'/'three'/'four')

Obtaining information about data types

We now understand how to retrieve information about the data types of columns.

Import the *subset-covid-data.csv* file and read the data into a DataFrame, as shown in the following.

CODE:

```
data=pd.read_csv('subset-covid-data.csv')
data.head()
```

Output:

	dateRep	day	month	year	cases	deaths	countriesAndTerritories	geoId	countryterritoryCode	popData2018	continentExp
0	2020-05-27	27	5	2020	658	1	Afghanistan	AF	AFG	37172386.0	Asia
1	2020-05-26	26	5	2020	591	1	Afghanistan	AF	AFG	37172386.0	Asia
2	2020-05-25	25	5	2020	584	2	Afghanistan	AF	AFG	37172386.0	Asia
3	2020-05-24	24	5	2020	782	11	Afghanistan	AF	AFG	37172386.0	Asia
4	2020-05-23	23	5	2020	540	12	Afghanistan	AF	AFG	37172386.0	Asia

Using the *dtypes* attribute, we can obtain the type of columns in this DataFrame.

CODE:

```
data.dtypes
```

Output:

```
country        object
continent      object
date           object
```

```
day                int64
month              int64
year               int64
cases              int64
deaths             int64
country_code      object
population       float64
dtype: object
```

As we discussed in the previous chapter, the kind of mathematical operations and graphs that can be used differ for categorical and continuous variables. Knowing the data types of columns helps us figure out how to analyze the variables. The columns that have the Pandas data type "object" or "category" are categorical variables, whereas variables with data types like "int64" and "float64" are continuous.

Get the count of each data type

To obtain the number of columns belonging to each data type, we use the *get_dtype_counts* method:

CODE:

```
data.get_dtype_counts()
```

Output:

```
float64    1
int64      5
object     4
dtype: int64
```

Select particular data types

Using the *select_dtypes* method, we can filter the columns based on the type of data you want to select:

CODE:

```
data.select_dtypes(include='number').head()
```

```
#This will select all columns that have integer and floating-point data and
exclude the rest. The head parameter has been used to limit the number of
records being displayed.
```

Output:

	day	month	year	cases	deaths	population
0	12	4	2020	34	3	37172386.0
1	12	4	2020	17	0	2866376.0
2	12	4	2020	64	19	42228429.0
3	12	4	2020	21	2	77006.0
4	12	4	2020	0	0	30809762.0

Calculating the memory usage and changing data types of columns

We can find the memory usage (in bytes) of a Series or a DataFrame by using the *memory_usage* method. We include the *deep* parameter while using this method to get a more comprehensive picture of the memory usage at the system level.

CODE:

```
data['continent'].memory_usage(deep=True)
```

Output:

```
13030
```

Let us see if we can reduce the memory usage of this column. First, let us find its current data type.

CODE:

```
data['continent'].dtype
```

Output:

```
dtype('O')
```

As we can see, this column occupies 13030 bytes of memory and has a data type of "O". The Pandas categorical data type is useful for storing qualitative variables that have only a few unique values, as this reduces memory usage. Since the continent column has only a few unique values ("Europe", "Asia", "America", "Africa", "Oceania"), let us change the data type of this column from *object* to *categorical,* and see if this reduces memory usage. We use the *astype* method for changing data types.

CODE:

```
data['continent']=data['continent'].astype('category')
data['continent'].memory_usage(deep=True)
```

Output:

```
823
```

The memory usage seems to have reduced quite a bit after changing the data type. Let us calculate the exact percentage reduction.

CODE:

```
(13030-823)/13030
```

Output:

```
0.936838066001535
```

A significant reduction in memory usage, around 93%, has been achieved by changing the data type from object to categorical.

Indexers and selection of subsets of data

In Pandas, there are many ways of selecting and accessing data, as listed in the following.

- *loc* and *iloc* indexers
- *ix* indexer
- *at* and *iat* indexers
- indexing operator []

The preferred method for data retrieval is through the use of the *loc* and *iloc* indexers. Both indexers and the indexing operator enable access to an object using indexes. Note that an indexer is different from the indexing operator, which is a pair of square brackets containing the index. While we have used the indexing operator [], for selecting data from objects like lists, tuples, and NumPy, the use of this operator is not recommended.

For instance, if we want to select the first row in Pandas, we would use the first statement given in the following.

CODE:

```
data.iloc[0] #correct
data[0] #incorrect
```

Understanding loc and iloc indexers

The **loc indexer** works by selecting data using index labels, which is similar to how data is selected in dictionaries in Python, using keys associated with values.

The **iloc indexer,** on the other hand, selects data using the integer location, which is similar to how individual elements are in lists and arrays.

Note that *loc* and *iloc* being indexers, and are followed by square brackets, not round brackets (like in the case of functions or methods). The index values before the comma refer to the row indexes, and the index values after the comma refer to the column indexes.

Let us consider some examples to understand how the *loc* and *iloc* indexers work. We again use the covid-19 dataset ("subset-covid-data.csv") for these examples.

CODE:

```
data=pd.read_csv('subset-covid-data.csv',index_col='date')
```

Here, we are using the column 'date' as the index.

Selecting consecutive rows

We can use *iloc* for this, since we know the index (first five) of the rows to be retrieved:

CODE:

```
data.iloc[0:5]
```

Output:

	country	continent	day	month	year	cases	deaths	country_code	population
date									
2020-04-12	Afghanistan	Asia	12	4	2020	34	3	AFG	37172386.0
2020-04-12	Albania	Europe	12	4	2020	17	0	ALB	2866376.0
2020-04-12	Algeria	Africa	12	4	2020	64	19	DZA	42228429.0
2020-04-12	Andorra	Europe	12	4	2020	21	2	AND	77006.0
2020-04-12	Angola	Africa	12	4	2020	0	0	AGO	30809762.0

Note that we mention only the row indexes, and in the absence of a column index, all the columns are selected by default.

Selecting consecutive columns

We can use *iloc* for this since the index values (0,1,2) for the first three columns are known.

CODE:

```
data.iloc[:,:3]
```

Or

```
data.iloc[:,0:3]
```

Output (only first five rows shown)

	country	continent	day
date			
2020-04-12	Afghanistan	Asia	12
2020-04-12	Albania	Europe	12
2020-04-12	Algeria	Africa	12
2020-04-12	Andorra	Europe	12
2020-04-12	Angola	Africa	12

While we can skip the column indexes, we cannot skip the row indexes.

The following syntax would not work:

CODE:

```
data.iloc[,0:3] #incorrect
```

In this example, we are selecting all the rows and three columns. On either side of the colon (:) symbol, we have a start and a stop value. If both start and stop values are missing, it means all values are to be selected. If the starting index is missing, it assumes a default value of 0. If the stop index value is missing, it assumes the last possible positional value of the index (one minus the number of columns or rows).

Selecting a single row

Let us select the 100th row using the *iloc* indexer. The 100th row has an index of 99 (since the index numbering starts from 0).

CODE:

```
data.iloc[99]
```

Output:

```
country              Jamaica
continent            America
day                       12
month                      4
year                    2020
cases                      4
deaths                     0
country_code             JAM
population       2.93486e+06
Name: 2020-04-12, dtype: object
```

Selecting rows using their index labels

Select the rows with the date as 2020-04-12. Here, we use the loc indexer since we know the index labels for the rows that need to be selected but do not know their position.

```
data.loc['2020-04-12']
```

Output (only first five rows shown):

	country	continent	day	month	year	cases	deaths	country_code	population
date									
2020-04-12	Afghanistan	Asia	12	4	2020	34	3	AFG	37172386.0
2020-04-12	Albania	Europe	12	4	2020	17	0	ALB	2866376.0
2020-04-12	Algeria	Africa	12	4	2020	64	19	DZA	42228429.0
2020-04-12	Andorra	Europe	12	4	2020	21	2	AND	77006.0
2020-04-12	Angola	Africa	12	4	2020	0	0	AGO	30809762.0

Selecting columns using their name

Let us select the column named "cases". Since the name of a column acts as its index label, we can use the loc indexer.

CODE:

```
data.loc[:,'cases']
```

Output:

```
date
2020-04-12        34
2020-04-12        17
2020-04-12        64
2020-04-12        21
2020-04-12         0
```

Using negative index values for selection

Let us select the first five rows and last three columns. Here, we are using negative indices to select the last three columns. The last column has a positional index value of -1, the last but one column has an index value of -2, and so on. The step size is -1. We skip the start value for the row slice (:5) since the default value is 0.

CODE:

```
data.iloc[:5,-1:-4:-1]
```

Output:

date	population	country_code	deaths
2020-04-12	37172386.0	AFG	3
2020-04-12	2866376.0	ALB	0
2020-04-12	42228429.0	DZA	19
2020-04-12	77006.0	AND	2
2020-04-12	30809762.0	AGO	0

Selecting nonconsecutive rows and columns

To select a set of rows or columns that are not consecutive, we need to enclose the rows or column index positions or labels in a list. In the following example, we select the second and fifth row, along with the first and fourth columns.

CODE:

```
data.iloc[[1,5],[0,3]]
```

Output:

date	country	month
2020-04-12	Albania	4
2020-04-12	Anguilla	4

Other (less commonly used) indexers for data access

The reason the indexers *loc* and *iloc* are recommended for slicing or selecting subsets of data from Series and DataFrame objects is that they have clear rules to select data either exclusively by their labels (in case of loc) or through their position (in case of iloc). However, it is essential to understand the other indexers supported in Pandas, explained in the following section.

ix indexer

The *ix* indexer allows us to select data by combining the index label and the location. This method of selection is in contrast to the method used by the *loc* and *iloc* indexers, which do not allow us to mix up the position and the label. With the *ix* indexer not having a clear rule to select data, there is room for much ambiguity, and this indexer has now been deprecated (which means that although it is still supported, it should not be used). For demonstration purposes, let us see how the ix indexer works. Let us select the first five rows of the column `cases' in our dataset.

CODE:

```
data.ix[:5,'cases']
```

Output:

```
 C:\Users\RA\Anaconda3\lib\site-packages\ipykernel_launcher.py:1:
DeprecationWarning:
.ix is deprecated. Please use
.loc for label based indexing or
.iloc for positional indexing

See the documentation here:
http://pandas.pydata.org/pandas-docs/stable/indexing.html#ix-indexer-is-
deprecated
  """Entry point for launching an IPython kernel.

date
2020-04-12    34
2020-04-12    17
2020-04-12    64
```

```
2020-04-12    21
2020-04-12     0
Name: cases, dtype: int64
```

Note that the use of the ix indicator leads to a warning asking the user to use either loc or iloc in its place.

The indexing operator - [ ]

Even though the indexing operator is not the preferred mode for data selection or slicing in Pandas, it still has its uses. One appropriate application of this operator is for selecting columns from a DataFrame. The argument is the name of the column that is mentioned as a string (enclosed within quotes).

For instance, the following statement would select the population column from our COVID dataset.

CODE:

```
data['population']
```

Output (only first five rows shown):

```
date
2020-04-12      37172386.0
2020-04-12       2866376.0
2020-04-12      42228429.0
2020-04-12         77006.0
2020-04-12      30809762.0
```

To select multiple columns, we pass the column names as strings within a list, as shown in the example in the following:

CODE:

```
data[['country','population']]
```

Output (truncated):

	countriesAndTerritories	popData2018
0	Afghanistan	37172386.0
1	Afghanistan	37172386.0
2	Afghanistan	37172386.0
3	Afghanistan	37172386.0
4	Afghanistan	37172386.0

The indexing operator can also be used for selecting a set of consecutive rows.

CODE:

```
data[:3]
```

Output:

	day	month	year	cases	deaths	countriesAndTerritories	geoId	countryterritoryCode	popData2018	continentExp
dateRep										
2020-05-27	27	5	2020	658	1	Afghanistan	AF	AFG	37172386.0	Asia
2020-05-26	26	5	2020	591	1	Afghanistan	AF	AFG	37172386.0	Asia
2020-05-25	25	5	2020	584	2	Afghanistan	AF	AFG	37172386.0	Asia

However, it cannot be used to select a series of nonconsecutive rows, as this will raise an error. The following statement would not work.

CODE:

```
data[[3,5]] #incorrect
```

Another limitation of the indexing operator is that it cannot be used to select rows and columns simultaneously. The following statement would also not work.

CODE:

```
data[:,3] #incorrect
```

at and iat indexers

There are two other less commonly used indexers - *at* (similar to *loc*, works with labels) and *iat* (similar to *iloc*, works with positions). The three main features of the *at* and *iat* indexers are

- They can be used only for selecting scalar (single) values from a Series or DataFrame.
- Both row and column indexes need to be supplied as arguments to these indexers since they return a single value. We cannot obtain a set of rows or columns with this indexer, which is possible with the other indexers.
- These indexers are quicker at retrieving data than loc and iloc.

Let us understand how these indexers work with the help of an example.

Import the subset-covid-data.csv dataset.

```
data=pd.read_csv('subset-covid-data.csv')
```

The *at* indexer works just like loc, and you need to pass the row index label and the column name as arguments.

Let us try to retrieve the population value in the first row. Since we have not set an index for this DataFrame, the index labels and positions would be the same.

CODE:

```
data.at[0,'population']
#0 is the index label as well as the position
```

Output:

```
37172386.0
```

The *iat* indexer is similar to the *iloc* indexer, with the row/column indexes being passed as arguments.

CODE:

```
data.iat[0,9]
#0,9 is the position of the first record of the population column
```

The output is the same as the one for the previous statement.

Boolean indexing for selecting subsets of data

In the previous examples that we looked at, we used various indexers to retrieve data based on the position or label. With Boolean indexing, we use conditional statements to filter data based on their values. A single condition may be specified, or multiple conditions can be combined using the bitwise operators - & (and), | (or), ~ (not).

Let us consider an example to understand this. Here, we select all records where the name of the continent is "Asia", and the country name starts with the letter "C".

CODE:

```
data[(data['continent']=='Asia') & (data['country'].str.startswith('C'))]
```

Output:

	country	continent	date	day	month	year	cases	deaths	country_code	population
33	Cambodia	Asia	2020-04-12	12	4	2020	2	0	KHM	1.624980e+07
42	China	Asia	2020-04-12	12	4	2020	93	0	CHN	1.392730e+09

Using the query method to retrieve data

While we combine multiple conditions as in the previous example, the readability of the code may suffer. The *query* method can be used in such cases.

Let us retrieve all the records where the name of the continent is "Asia" and the number of cases is higher than 500. Note the syntax where we enclose each condition within double quotes and use the logical *and* operator, instead of the bitwise operator, &.

CODE:

```
data.query("(continent=='Asia')""and (cases>=500)")
```

Output:

	country	continent	date	day	month	year	cases	deaths	country_code	population
91	India	Asia	2020-04-12	12	4	2020	909	34	IND	1.352617e+09
93	Iran	Asia	2020-04-12	12	4	2020	1837	125	IRN	8.180027e+07
100	Japan	Asia	2020-04-12	12	4	2020	1401	10	JPN	1.265291e+08
190	Turkey	Asia	2020-04-12	12	4	2020	5138	95	TUR	8.231972e+07

Further reading

See more on:

- Query method: `https://pandas.pydata.org/pandas-docs/stable/reference/api/pandas.DataFrame.query.html`
- Indexing in Pandas: `https://pandas.pydata.org/docs/user_guide/indexing.html`

Operators in Pandas

Pandas uses the following operators that can be applied to a whole series. While Python would require a loop to iterate through every element in a list or dictionary, Pandas takes advantage of the feature of vectorization implemented in NumPy that enables these operators to be applied on every element in a sequence, eliminating the need for iteration and loops. The different types of operators are listed in Table 6-7.

Table 6-7. *Pandas Operators*

Type of operator	Operators included
Arithmetic Operators	+addition), -(subtraction), *(multiplication),**(power),%(remainder operator),/(division),//(floor division, for getting the quotient). The functions performed by arithmetic operators can be replicated using the following methods: add for +, sub for -, mul for *, div for /, mod for %, and pow for **.
Comparison Operators	== (equality),<(less than),>(greater than),<=(less than or equal to),>=(greater than or equal to),!=(not equal to)
Logical Operators	&,|,~. Pandas, like NumPy, uses the bitwise operators (&,|,~) as logical operators, as these operators operate on every element of a Series. Note that these operators are different from the logical operators used in Python, where the keywords *and*, *or*, and *not* are used.

Representing dates and times in Pandas

In Pandas, there is a single Timestamp function that can be used to define a date, time, or a combination of a date and a time. This is in contrast to the implementation in Python, which requires separate objects to define a date or time. The *pd.Timestamp* function is equivalent to the following functions in Python: *datetime.date, datetime.time, datetime.datetime.*

As an example, let us represent the date 25th December 2000 in Pandas using the *pd. Timestamp* function.

CODE:

```
pd.Timestamp('25/12/2000')
```

Output:

```
Timestamp('2000-12-25 00:00:00')
```

The *Timestamp* function is very flexible and accepts parameters in a variety of formats. The preceding output can also be replicated using any of the following statements.

```
#different input formats for creating a Timestamp object

pd.Timestamp('25 December 2000')

pd.Timestamp('December 25 2000')

pd.Timestamp('12/25/2000')

pd.Timestamp('25-12-2000')

pd.Timestamp(year=2000,month=12,day=25)

pd.Timestamp('25-12-2000 12 PM')

pd.Timestamp('25-12-2000 0:0.0')
```

The *pd.Timestamp* function helps us define a date, time, and a combination of these two. However, this function does not work if we need to define a duration of time. A separate function, *pd.Timedelta,* helps us create objects that store a time duration. This is equivalent to the *datetime.timedelta* function in Python.

Let us define a duration of time in Pandas using the *Timedelta* function.

CODE:

```
pd.Timedelta('45 days 9 minutes')
```

Output:

```
Timedelta('45 days 00:09:00')
```

Like the *Timestamp* function, the *Timedelta* function is flexible in what it accepts as input parameters. The preceding statement can also be written as follows.

CODE:

```
pd.Timedelta(days=45,minutes=9)
```

We can also add the *unit* parameter to create a *Timedelta* object. In the following line of code, the parameter unit with the value 'm' denotes minutes, and we add 500 minutes to the base time of 00:00:00 hours.

CODE:

```
pd.Timedelta(500,unit='s')
```

Output:

```
Timedelta('0 days 08:20:00')
```

Converting strings into Pandas Timestamp objects

Dates are generally represented as strings and need to be converted to a type that can be understood by Pandas. The *pd.to_datetime* function converts the date to a Timestamp object. Converting it to this format helps with comparing two dates, adding or subtracting a time duration from a given date, and extracting individual components (like day, month, and year) from a given date. It also helps with representing dates that are not in the traditional "day-month-year" or "month-day-year" format.

Let us consider an example to understand this. Consider the date represented as a string "11:20 AM, 2nd April 1990". We can convert this into a Timestamp object and specify the format parameter so that the individual components like the day, month, and year are parsed correctly. The *format* parameter in the *pd.to_datetime* function with its formatting codes (like %H, %M), helps with specifying the format in which this date is written. %H represents the hour, %M represents the minutes, %d is for the day, %m is for the month, and %Y is for the year.

CODE:

```
a=pd.to_datetime('11:20,02/04/1990', format='%H:%M,%d/%m/%Y')
a
```

Output:

```
Timestamp('1990-04-02 11:20:00')
```

Now that this date has been converted into a *Timestamp* object, we can perform operations on it. A *Timedelta* object can be added to a Timestamp object.

Let us add four days to this date:

CODE:

```
a+pd.Timedelta(4,unit='d')
```

Output:

```
Timestamp('1990-04-06 11:20:00')
```

Extracting the components of a Timestamp object

Once the date is converted to a Pandas Timestamp object using the *pd.to_datetime* function, the individual components of the date variable can be extracted using the relevant attributes.

CODE:

```
#extracting the month
a.month
```

Output:

```
4
```

CODE:

```
#extracting the year
a.year
```

Output:

```
1990
```

CODE:

```
#extracting the day
a.day
```

Output:

```
2
```

We can also use the minute and hour attribute to extract the minutes and hour from the date.

Further reading

Learn more about the Pandas Timestamp function: https://pandas.pydata.org/pandas-docs/stable/reference/api/pandas.Timestamp.html

Grouping and aggregation

Aggregation is the process of summarizing a group of values into a single value.

Hadley Wickham, a statistician, laid down the "Split-Apply-Combine" methodology (the paper can be accessed here: https://www.jstatsoft.org/article/view/v040i01/v40i01.pdf), which has three steps:

1. Split the data into smaller groups that are manageable and independent of each other. This is done using the *groupby* method in Pandas.
2. Apply functions on each of these groups. We can apply any of the aggregation functions, including minimum, maximum, median, mean, sum, count, standard deviation, variance, and size. Each of these aggregate functions calculate the aggregate value of the entire group. Note that we can also write a customized aggregation function.
3. Combine the results after applying functions to each group into a single combined object.

In the following section, we look at the *groupby* method, aggregation functions, the *transform, filter,* and *apply* methods, and the properties of the *groupby* object.

Here, we again use the same COVID-19 dataset, which shows the number of cases and deaths for all countries on 12th April 2020.

CODE:

```
df=pd.read_csv('subset-covid-data.csv')
df.head()
```

Output:

	country	continent	date	day	month	year	cases	deaths	country_code	population
0	Afghanistan	Asia	2020-04-12	12	4	2020	34	3	AFG	37172386.0
1	Albania	Europe	2020-04-12	12	4	2020	17	0	ALB	2866376.0
2	Algeria	Africa	2020-04-12	12	4	2020	64	19	DZA	42228429.0
3	Andorra	Europe	2020-04-12	12	4	2020	21	2	AND	77006.0
4	Angola	Africa	2020-04-12	12	4	2020	0	0	AGO	30809762.0

As we can see, there are several countries belonging to the same continent. Let us find the total number of cases and deaths for each continent. For this, we need to do grouping using the 'continent' column.

CODE:

```
df.groupby('continent')['cases','deaths'].sum()
```

Output:

	cases	deaths
continent		
Africa	714	52
America	33519	2111
Asia	12979	383
Europe	34141	3571
Oceania	68	4
Other	0	0

Here, we are grouping by the column "continent", which becomes the *grouping column*. We are aggregating the values of the number of cases and deaths, which makes the columns named "cases" and "deaths" the *aggregating columns*. The sum method, which becomes our *aggregating function*, calculates the total of cases and deaths for all countries belonging to a given continent. Whenever you perform a *groupby* operation, it is recommended that these three elements (grouping column, aggregating column, and aggregating function) be identified at the outset.

The following thirteen aggregate functions can be applied to groups: *sum(), max(), min(), std(), var(), mean(), count(), size(), sem(), first(), last(), describe(), nth()*.

We can also use the *agg* method, with *np.sum* as an attribute, which produces the same output as the previous statement:

CODE:

```
df.groupby('continent')['cases','deaths'].agg(np.sum)
```

The *agg* method can accept any of the aggregating methods, like mean, sum, max, and so on, and these methods are implemented in NumPy.

We can also pass the aggregating column and the aggregating method as a dictionary to the *agg* method, as follows, which would again produce the same output.

CODE:

```
df.groupby('continent').agg({'cases':np.sum,'deaths':np.sum})
```

If there is more than one grouping column, use a list to save the column names as strings and pass this list as an argument to the *groupby* method.

Further reading on aggregate functions: `https://pandas.pydata.org/pandas-docs/stable/user_guide/groupby.html#aggregation`

Examining the properties of the groupby object

The result of applying the *groupby* method is a groupby object. This *groupby* object has several properties that are explained in this section.

Data type of *groupby* object

The data type of a groupby object can be accessed using the type function.

CODE:

```
grouped_continents=df.groupby('continent')
type(grouped_continents)
```

Output:

```
pandas.core.groupby.generic.DataFrameGroupBy
```

Each group of the groupby object is a separate DataFrame.

Obtaining the names of the groups

The *groupby* object has an attribute called *groups*. Using this attribute on the *groupby* object would return a dictionary, with the keys of this dictionary being the names of the groups.

CODE:

```
grouped_continents.groups.keys()
```

Output:

```
dict_keys(['Africa', 'America', 'Asia', 'Europe', 'Oceania', 'Other'])
```

Returning records with the same position in each group using the nth method

Let us say that you want to see the details of the fourth country belonging to each continent. Using the *nth* method, we can retrieve this data by using a positional index value of 3 for the fourth position.

CODE:

```
grouped_continents.nth(3)
```

Output:

continent	country	date	day	month	year	cases	deaths	country_code	population
Africa	Botswana	2020-04-12	12	4	2020	0	0	BWA	2254126.0
America	Aruba	2020-04-12	12	4	2020	6	0	ABW	105845.0
Asia	Bhutan	2020-04-12	12	4	2020	0	0	BTN	754394.0
Europe	Austria	2020-04-12	12	4	2020	247	18	AUT	8847037.0
Oceania	Guam	2020-04-12	12	4	2020	3	1	GUM	165768.0

Get all the data for a particular group using the get_group method

Use the *get_group* method with the name of the group as an argument to this method. In this example, we retrieve all data for the group named 'Europe.'

CODE:

```
grouped_continents.get_group('Europe')
```

Output (contains 54 records; only first four records shown in the following):

	country	date	day	month	year	cases	deaths	country_code	population
1	Albania	2020-04-12	12	4	2020	17	0	ALB	2866376.0
3	Andorra	2020-04-12	12	4	2020	21	2	AND	77006.0
8	Armenia	2020-04-12	12	4	2020	30	2	ARM	2951776.0
11	Austria	2020-04-12	12	4	2020	247	18	AUT	8847037.0

We have seen how to apply aggregate functions to the groupby object. Now let us look at some other functions, like filter, apply, and transform, that can also be used with a groupby object.

Filtering groups

The *filter* method removes or filters out groups based on a particular condition. While the agg (aggregate) method returns one value for each group, the filter method returns records from each group depending on whether the condition is satisfied.

Let us consider an example to understand this. We want to return all the rows for the continents where the average death rate is greater than 40. The *filter* method is called on a groupby object and the argument to the *filter* method is a lambda function or a predefined function. The *lambda* function here calculates the average death rate for every group, represented by the argument "x". This argument is a DataFrame representing each group (which is the continent in our example). If the condition is satisfied for the group, all its rows are returned. Otherwise, all the rows of the group are excluded.

CODE:

```
grouped_continents=df.groupby('continent')
grouped_continents.filter(lambda x:x['deaths'].mean()>=40)
```

Output (only first five rows shown):

	country	continent	date	day	month	year	cases	deaths	country_code	population
1	Albania	Europe	2020-04-12	12	4	2020	17	0	ALB	2866376.0
3	Andorra	Europe	2020-04-12	12	4	2020	21	2	AND	77006.0
5	Anguilla	America	2020-04-12	12	4	2020	0	0	NaN	NaN
6	Antigua_and_Barbuda	America	2020-04-12	12	4	2020	0	0	ATG	96286.0
7	Argentina	America	2020-04-12	12	4	2020	162	7	ARG	44494502.0
8	Armenia	Europe	2020-04-12	12	4	2020	30	2	ARM	2951776.0

In the output, we see that only the rows for the groups (continents) 'America' and 'Europe' are returned since these are the only groups that satisfy the condition (group mean death rate greater than 40).

Transform method and groupby

The *transform* method is another method that can be used with the *groupby* object, which applies a function on each value of the group. It returns an object that has the same rows as the original data frame or Series and is similarly indexed as well.

Let us use the *transform* method on the population column to obtain the population in millions by dividing each value in the row by 1000000.

CODE:

```
grouped_continents['population'].transform(lambda x:x/1000000)
```

Output (only first five rows and last two rows shown; actual output contains 206 rows):

```
0       37.172386
1        2.866376
2       42.228429
3        0.077006
4       30.809762
.
..
...
204     17.351822
205     14.439018
Name: population, Length: 206, dtype: float64
```

Notice that while the *filter* method returns lesser records as compared to its input object, the *transform* method returns the same number of records as the input object.

In the preceding example, we have applied the *transform* method on a Series. We can also use it on an entire DataFrame. A common application of the *transform* method is used to fill null values. Let us fill the missing values in our DataFrame with the value 0. In the output, notice that the values for the country code and population for the country 'Anguilla' (which were missing earlier) are now replaced with the value 0.

CODE:

```
grouped_continents.transform(lambda x:x.fillna(0))
```

Output:

	country	date	day	month	year	cases	deaths	country_code	population
0	Afghanistan	2020-04-12	12	4	2020	34	3	AFG	37172386.0
1	Albania	2020-04-12	12	4	2020	17	0	ALB	2866376.0
2	Algeria	2020-04-12	12	4	2020	64	19	DZA	42228429.0
3	Andorra	2020-04-12	12	4	2020	21	2	AND	77006.0
4	Angola	2020-04-12	12	4	2020	0	0	AGO	30809762.0
5	Anguilla	2020-04-12	12	4	2020	0	0	0	0.0

The *transform* method can be used with any Series or a DataFrame and not just with *groupby* objects. Creating a new column from an existing column is a common application of the *transform* method.

Apply method and groupby

The *apply* method "applies" a function to each group of the *groupby* object. The difference between the apply and transform method is that the apply method is more flexible in that it can return an object of any shape while the transform method needs to return an object of the same shape.

The *apply* method can return a single (scalar) value, Series or DataFrame, and the output need not be in the same structure as the input. Also, while the *transform* method applies the function on each column of a group, the *apply* method applies the function on the entire group.

Let us use the *apply* method to calculate the total missing values in each group (continent).

CODE:

```
grouped_continents.apply(lambda x:x.isna().sum())
```

Output:

	country	continent	date	day	month	year	cases	deaths	country_code	population
continent										
Africa	0	0	0	0	0	0	0	0	0	1
America	0	0	0	0	0	0	0	0	3	3
Asia	0	0	0	0	0	0	0	0	0	0
Europe	0	0	0	0	0	0	0	0	0	0
Oceania	0	0	0	0	0	0	0	0	0	0
Other	0	0	0	0	0	0	0	0	1	0

The *apply* method, similar to the transform method, can be used with Series and DataFrame objects in addition to the *groupby* object.

How to combine objects in Pandas

In Pandas, there are various functions to combine two or more objects, depending on whether we want to combine them horizontally or vertically. In this section, we cover the four methods used for combining objects - *append, join, concat,* and *merge.*

Append method for adding rows

This method is used to add rows to an existing DataFrame or Series object, but cannot be used to add columns. Let us look at this with an example:

Let us create the following DataFrame:

CODE:

```
periodic_table=pd.DataFrame({'Atomic Number':[1,2,3,4,5],'Element':
['Hydrogen','Helium','Lithium','Beryllium','Boron'],'Symbol':['H','He',
'Li','Be','B']})
```

We now add a new row (in the form of a dictionary object) by passing it as an argument to the *append* method.

We also need to remember to set the value of the *ignore_index* parameter as True. Setting it to "True" replaces the old index with a new index.

CODE:

```
periodic_table.append({'Atomic Number':6,'Element':'Carbon','Symbol':'C'},i
gnore_index=True)
```

Output:

	Atomic Number	Element	Symbol
0	1	Hydrogen	H
1	2	Helium	He
2	3	Lithium	Li
3	4	Beryllium	Be
4	5	Boron	B
5	6	Carbon	C

Note that if we skip the *ignore_index* parameter while using the append function, we will get an error, as shown in the following:

CODE:

```
periodic_table.append({'Atomic Number':6,'Element':'Carbon','Symbol':'C'})
```

Output:

```
---------------------------------------------------------------------------
TypeError                                 Traceback (most recent call last)
<ipython-input-164-2e1fc586027a> in <module>
----> 1 periodic_table.append({'Atomic Number':6,'Element':'Carbon','Symbol':'C'})

~\Anaconda3\lib\site-packages\pandas\core\frame.py in append(self, other, ignore_index, verify_integrity, sort)
   6656                 other = Series(other)
   6657             if other.name is None and not ignore_index:
-> 6658                 raise TypeError('Can only append a Series if ignore_index=True'
   6659                                 ' or if the Series has a name')
   6660 

TypeError: Can only append a Series if ignore_index=True or if the Series has a name
```

Using the *append* method, we can also add multiple rows by defining each row as a Series object and passing these Series objects as a list to the append method. The *pd.Series* method has a name attribute that assigns an index label to a Series.

CODE:

```
series1=pd.Series({'Atomic Number':7,'Element':'Carbon','Symbol':'C'},name=
len(periodic_table))
series2=pd.Series({'Atomic Number':8,'Element':'Oxygen','Symbol':'O'},name=
len(periodic_table)+1)
periodic_table.append([series1, series2])
```

Output:

	Atomic Number	Element	Symbol
0	1	Hydrogen	H
1	2	Helium	He
2	3	Lithium	Li
3	4	Beryllium	Be
4	5	Boron	B
5	7	Carbon	C
6	8	Oxygen	O

Note that we did not use the *ignore_index* parameter this time, since we have used the *name* parameter (refer to the error message shown earlier where it is mentioned that we can use either the *ignore_index* parameter or *name* parameter with the append method). Using the *name* parameter prevents the resetting of the index, which happens when we include the *ignore_index* parameter.

Understanding the various types of joins

Before we move on to the other methods for combining Pandas objects, we need to understand the concepts of an inner, outer, left, and right join. When you join two objects, the type of join determines which records from these objects get included in the final result set.

- Left join: All rows from the object on the left included in the combined object. Rows from the object on the right that match those from the left included.
- Right join: All rows from the object on the right included in the combined object. Rows from the object on the left that match those from the left included.
- Outer join: All rows from both objects included in the combined object (whether they match or not).
- Inner join: Only matching rows from both objects included.

Concat function (adding rows or columns from other objects)

This function gives us the option to add both rows and columns to a Pandas object. By default, it works on the row axis and adds rows.

Let us look at how the *concat* function works with an example. Here, we join two DataFrame objects vertically. The second DataFrame object is added after the last row of the first DataFrame object.

CODE:

```
periodic_table=pd.DataFrame({'Atomic Number':[1,2,3,4,5],'Element':
['Hydrogen','Helium','Lithium','Beryllium','Boron'],'Symbol':['H','He',
'Li','Be','B']})
```

```
periodic_table_extended=pd.DataFrame({'Atomic Number':[8,9,10],'Element':['
Oxygen','Fluorine','Neon'],'Symbol':['O','F','Ne']})
#Join these two DataFrames just created vertically using the concat
function:
pd.concat([periodic_table,periodic_table_extended])
```

Output:

	Atomic Number	Element	Symbol
0	1	Hydrogen	H
1	2	Helium	He
2	3	Lithium	Li
3	4	Beryllium	Be
4	5	Boron	B
0	8	Oxygen	O
1	9	Fluorine	F
2	10	Neon	Ne

We can also concatenate objects side-by-side along the column axis, as shown in the following.

CODE:

```
pd.concat([periodic_table,periodic_table_extended],axis=1)
```

Output:

	Atomic Number	Element	Symbol	Atomic Number	Element	Symbol
0	1	Hydrogen	H	8.0	Oxygen	O
1	2	Helium	He	9.0	Fluorine	F
2	3	Lithium	Li	10.0	Neon	Ne
3	4	Beryllium	Be	NaN	NaN	NaN
4	5	Boron	B	NaN	NaN	NaN

By default, the *concat* function performs an outer join, which returns all records of both the objects. The concatenated result set will have five records (equal to the length of the longer object – the first DataFrame). Since the second DataFrame has only three rows, you can see null values for the fourth and fifth row in the final concatenated object.

We can change this to an inner join by adding the *join* parameter. By using an inner join as shown in the following, the final result set with contain only those records from both the objects where the indices match.

CODE:

```
pd.concat([periodic_table,periodic_table_extended],axis=1,join='inner')
```

Output:

	Atomic Number	Element	Symbol	Atomic Number	Element	Symbol
0	1	Hydrogen	H	8	Oxygen	O
1	2	Helium	He	9	Fluorine	F
2	3	Lithium	Li	10	Neon	Ne

We can use the *keys* parameter to identify each of the objects that are being concatenated, in the final result set.

CODE:

```
pd.concat([periodic_table,periodic_table_extended],axis=1,keys=['1st
periodic table','2nd periodic table'])
```

Output:

	1st periodic table			2nd periodic table		
	Atomic Number	Element	Symbol	Atomic Number	Element	Symbol
0	1	Hydrogen	H	8.0	Oxygen	O
1	2	Helium	He	9.0	Fluorine	F
2	3	Lithium	Li	10.0	Neon	Ne
3	4	Beryllium	Be	NaN	NaN	NaN
4	5	Boron	B	NaN	NaN	NaN

Join method – index to index

The *join* method aligns two Pandas objects based on common index values. That is, it looks for matching index values in both objects and then align them vertically. The default type of join for this method is a left join.

Let us consider the following example where we join two objects.

CODE:

```
periodic_table.join(periodic_table_extended,lsuffix='_left',rsuffix='_right')
```

Output:

	Atomic Number_left	Element_left	Symbol_left	Atomic Number_right	Element_right	Symbol_right
0	1	Hydrogen	H	8.0	Oxygen	O
1	2	Helium	He	9.0	Fluorine	F
2	3	Lithium	Li	10.0	Neon	Ne
3	4	Beryllium	Be	NaN	NaN	NaN
4	5	Boron	B	NaN	NaN	NaN

Since the two DataFrame objects have common column names in the preceding example, we need to use the *lsuffix* and *rsuffix* parameters to differentiate between them. The indexes 0, 1, and 2 are common to both objects. The result set includes all the rows in the first DataFrame, and if there are rows with indices not matching in the second DataFrame, the value in all these rows is a null (denoted by NaN). The default join type used for the join method is a left join.

Merge method – SQL type join based on common columns

Like the *join* method, the *merge* method is also used to join objects horizontally. It is used when we join two DataFrame objects with a common column name. The main difference between the *join* and *merge* methods is that the *join* method combines the objects based on common index values, while the *merge* method combines the objects based on common column names. Another difference is that the default join type in the merge method is an inner join, while the join method performs a left join of the objects by default.

Let us look at how the merge method works with an example. The two DataFrame objects defined here have a column name that is common - Atomic Number. This is a scenario where we can apply the merge method.

CODE:

```
periodic_table=pd.DataFrame({'Atomic Number':[1,2,3,4,5],'Element':
['Hydrogen','Helium','Lithium','Beryllium','Boron'],'Symbol':['H','He',
'Li','Be','B']})
periodic_table_extended=pd.DataFrame({'Atomic Number':[1,2,3],'Natural':
'Yes'})
periodic_table.merge(periodic_table_extended)
```

Output:

	Atomic Number	Element	Symbol	Natural
0	1	Hydrogen	H	Yes
1	2	Helium	He	Yes
2	3	Lithium	Li	Yes

The presence of a common column name is essential for a merge operation, otherwise we would get an error. If there is a more than one column common between the two DataFrames, we can mention the column on which the merge is to be performed using the on parameter.

We can change the default join type (which is an inner join) using the *how* parameter.

CODE:

```
periodic_table.merge(periodic_table_extended,how='outer')
```

Output:

	Atomic Number	Element	Symbol	Natural
0	1	Hydrogen	H	Yes
1	2	Helium	He	Yes
2	3	Lithium	Li	Yes
3	4	Beryllium	Be	NaN
4	5	Boron	B	NaN

If we have the same column in both the objects that are being joined but their names are different, we can use parameters in the merge method to differentiate these columns.

In the following example, there are two columns with the same values but different names. In the first DataFrame object, the name of the column is 'Atomic Number', while in the second DataFrame object, the name of the column is 'Number'.

CODE:

```
periodic_table=pd.DataFrame({'Atomic Number':[1,2,3,4,5],'Element':
['Hydrogen','Helium','Lithium','Beryllium','Boron'],'Symbol':['H','He',
'Li','Be','B']})
periodic_table_extended=pd.DataFrame({'Number':[1,2,3],'Natural':'Yes'})
```

Using the *left_on* parameter for the column on the left, and *right_on* parameter for the column on the right, we merge the two objects as follows:

CODE:

```
periodic_table.merge(periodic_table_extended,left_on='Atomic Number',
right_on='Number')
```

Output:

	Atomic Number	Element	Symbol	Number	Natural
0	1	Hydrogen	H	1	Yes
1	2	Helium	He	2	Yes
2	3	Lithium	Li	3	Yes

Note that *append, merge,* and *join* are all DataFrame methods used with DataFrame objects while *concat* is a Pandas function.

Further reading:

Merge method: `https://pandas.pydata.org/pandas-docs/stable/reference/api/pandas.DataFrame.merge.html`

Join method: `https://pandas.pydata.org/pandas-docs/stable/reference/api/pandas.DataFrame.join.html#pandas.DataFrame.join`

Concat function: `https://pandas.pydata.org/pandas-docs/stable/reference/api/pandas.concat.html`

Append method: `https://pandas.pydata.org/pandas-docs/stable/reference/api/pandas.DataFrame.append.html`

Restructuring data and dealing with anomalies

As we have discussed earlier, data in its raw state is often messy and unfit for analysis. Most datasets require extensive treatment before they become fit for analysis and visualization. The most common problems encountered in datasets are given in the following.

- Data has missing values.
- Names of columns are not comprehensible.
- Variables are stored in rows as well as columns.
- A column may represent more than one variable.
- There are different observational units in a single table.
- There is data duplication.

- Data is structured around the wrong axis (for instance, horizontally instead of vertically).
- Variables are categorical, but we need them to be in a numeric format for performing calculations and visualizing them.
- The data type of a variable is not correctly recognized.
- Data contains spaces, commas, special symbols, and so on which need to be removed.

In the following sections, we understand how to handle missing and duplicate data, convert data from wide to long format, and how to use various methods like *pivot*, *stack*, and *melt*.

Dealing with missing data

Missing data in Pandas is represented by the value *NaN* (*Not a Number*), denoted as the keyword *np.nan*. We can use the *isna* or *isnull* method for finding null values. Both methods return a True (for a missing value) or False (for all other values) Boolean value for each object in the Series or DataFrame.

Let us see how many null values there are in the rainfall dataset.

CODE:

```
df=pd.read_csv('subset-covid-data.csv')
df.isna().sum().sum()
```

Output:

8

There are eight null values in this dataset. The *sum* method is used twice. The first sum method calculates the total number of missing values for each column, and the second sum method adds up these values to give the number of missing values in the entire DataFrame.

We have two options for dealing with this missing data - either we get rid of these values (drop them), or we substitute these values with a suitable measure (like the mean, median, or mode) that can be used as an approximation for the missing value. Let us look at each of these methods.

Dropping the missing data

The *dropna* method removes all the missing values in a DataFrame or Series.

CODE:

```
df.dropna()
```

Note that this method creates a copy of the data frame and does not modify the original DataFrame. To modify the original DataFrame, we use the inplace=True parameter.

CODE:

```
df.dropna(inplace=True)
```

Imputation

Imputation is the process of replacing missing values. In Pandas, there is an option to substitute the missing values with measures of central tendencies like the mean, median, or mode, using the *fillna* method. You can also fill the missing values with a fixed or constant value like 0.

We can use the forward fill technique to fill the missing value with the value just before it, or the backward fill technique to substitute the null value with the value just after it.

Using the same dataset (subset-covid-data.csv), let us try to understand the concepts of forward fill and backward fill.

CODE:

```
data=pd.read_csv('subset-covid-data.csv')
df=data[4:7]
df
```

As we can see, the DataFrame object, df (created from the original dataset), has missing values.

	country	continent	date	day	month	year	cases	deaths	country_code	population
4	Angola	Africa	2020-04-12	12	4	2020	0	0	AGO	30809762.0
5	Anguilla	America	2020-04-12	12	4	2020	0	0	NaN	NaN
6	Antigua_and_Barbuda	America	2020-04-12	12	4	2020	0	0	ATG	96286.0

Let us substitute the *NaN* value for the country `Anguilla' with the values preceding it (in the previous row), using the forward fill technique (*ffill*), as shown in the following.

CODE:

```
df.fillna(method='ffill')
```

Output:

	country	continent	date	day	month	year	cases	deaths	country_code	population
4	Angola	Africa	2020-04-12	12	4	2020	0	0	AGO	30809762.0
5	Anguilla	America	2020-04-12	12	4	2020	0	0	AGO	30809762.0
6	Antigua_and_Barbuda	America	2020-04-12	12	4	2020	0	0	ATG	96286.0

As we can see, the population field for Anguilla is substituted with the corresponding value of Angola (the record that precedes it).

We can also substitute the missing values using the backward fill technique, "bfill", which substitutes the missing values with the next valid observation that occurs in the row following it.

CODE:

```
df.fillna(method='bfill')
```

Output:

	country	continent	date	day	month	year	cases	deaths	country_code	population
4	Angola	Africa	2020-04-12	12	4	2020	0	0	AGO	30809762.0
5	Anguilla	America	2020-04-12	12	4	2020	0	0	ATG	96286.0
6	Antigua_and_Barbuda	America	2020-04-12	12	4	2020	0	0	ATG	96286.0

The missing population value for Anguilla is now substituted with the corresponding value from the next row (that of Antigua_and_Barbuda).

A standard method for imputation of missing values is to substitute the null values with the mean value of the other valid observations. The *fillna* method can be used for this purpose, as well.

Here, we are substituting the missing value in each column with the mean of the other two values in the column.

CODE:

```
df.fillna(df.mean())
```

Output:

	country	continent	date	day	month	year	cases	deaths	country_code	population
4	Angola	Africa	2020-04-12	12	4	2020	0	0	AGO	30809762.0
5	Anguilla	America	2020-04-12	12	4	2020	0	0	NaN	15453024.0
6	Antigua_and_Barbuda	America	2020-04-12	12	4	2020	0	0	ATG	96286.0

The missing population value for Anguilla is now substituted with the mean of the population figures for the other two countries (Angola &Antigua_and_Barbuda) in the DataFrame object, df.

Interpolation is another technique for estimating missing values in numeric columns, with the most straightforward interpolation technique being the linear interpolation method. In linear interpolation, the equation of a straight line is used to estimate unknown values from known values. If there are two points, (x_0, y_0) and (x_1,y_1), then an unknown point (x,y) can be estimated using the following equation:

$$y - y_0 = \left(\frac{y_1 - y_0}{x_1 - x_0}\right)(x - x_0)$$

In Pandas, there is an *interpolate* method that estimates an unknown value from known values.

CODE:

```
df.interpolate(method='linear')
```

Output:

	country	continent	date	day	month	year	cases	deaths	country_code	population
4	Angola	Africa	2020-04-12	12	4	2020	0	0	AGO	30809762.0
5	Anguilla	America	2020-04-12	12	4	2020	0	0	NaN	15453024.0
6	Antigua_and_Barbuda	America	2020-04-12	12	4	2020	0	0	ATG	96286.0

The missing values in each column are interpolated using the other values in the column.

Further reading: See more on missing data in Pandas: `https://pandas.pydata.org/pandas-docs/stable/user_guide/missing_data.html`

Data duplication

Redundancy in data is a common occurrence with many records containing the same data.

Let us consider the following DataFrame:

CODE:

```
periodic_table=pd.DataFrame({'Atomic Number':[1,2,3,4,5,5],'Element':['Hydr
ogen','Helium','Lithium','Beryllium','Boron','Boron'],'Symbol':['H','He','L
i','Be','B','B']})
```

	Atomic Number	Element	Symbol
0	1	Hydrogen	H
1	2	Helium	He
2	3	Lithium	Li
3	4	Beryllium	Be
4	5	Boron	B
5	5	Boron	B

As we can see, there are duplicates in this data (the last two rows are the same).

In Pandas, the presence of duplicates can be detected using the *duplicated* method. The *duplicated* method returns a Boolean value for each row, as shown in the following. Since the fifth row is the duplicate of the fourth row, the Boolean value is True.

CODE:

```
periodic_table.duplicated()
```

Output:

```
0    False
1    False
2    False
3    False
```

```
4    False
5     True
dtype: bool
```

By chaining this method with the *sum* method, we can find the total number of duplicates in the DataFrame.

CODE:

```
periodic_table.duplicated().sum()
```

Output:

```
1
```

Let us now get rid of the duplicates using the *drop_duplicates* method.

CODE:

```
periodic_table.drop_duplicates(inplace=True)
```

Output:

	Atomic Number	Element	Symbol
0	1	Hydrogen	H
1	2	Helium	He
2	3	Lithium	Li
3	4	Beryllium	Be
4	5	Boron	B

The duplicate row has been removed. Since the *drop_duplicates* method does not make changes to the actual DataFrame object, we need to use the *inplace* parameter.

By default, the *drop_duplicates* method keeps the first row among all the duplicate rows. If we want to keep the last row, we can use the parameter *keep='last'*, as shown in the following.

CODE:

```
periodic_table.drop_duplicates(keep='last')
```

Output:

	Atomic Number	Element	Symbol
0	1	Hydrogen	H
1	2	Helium	He
2	3	Lithium	Li
3	4	Beryllium	Be
5	5	Boron	B

Apart from dealing with redundant or missing data, we may need to replace data that does not add value to our analysis, like blank spaces or special characters. Removal or replacement of values can be done using the replace method, discussed earlier. We may also need to change the data types of the columns since Pandas may not recognize all the data types correctly, which can be done using the "astype" method.

Tidy data and techniques for restructuring data

Tidy data is a term developed by Hadley Wickham. According to a paper authored by him (Link: `http://vita.had.co.nz/papers/tidy-data.pdf`), these are the three principles of tidy data:

1. Columns correspond to variables in the data, and each variable maps to a single column.
2. The rows contain only observations, not variables.
3. Each data structure or table contains only one observational unit.

Note that making data tidy is different from data cleansing. Data cleansing is concerned with dealing with missing values and redundant information, removing filler characters, and changing inaccurate data types. On the other hand, converting data to a tidy format involves restructuring the data and arranging it along the right axis, to facilitate ease of analysis.

Let us understand this with an example, using the following DataFrame.

	Attributes	Anita	Binni	Chaya	Dinara
0	Age	10	7	8	9
1	Height	122	135	142	129

The preceding DataFrame displays the age (in years) and height (in centimeters) of four students. Though this data is readable, it is not in the "tidy" form. There are three issues with this DataFrame that go against the principles of tidy data:

- The names of the students cannot be used as column names. Instead, we need to have a single variable for all the names.
- The attributes "Age" and "Height" should not be used as observations in rows. They are in fact separate variables and should be individual columns.
- There are two different observational units in the same DataFrame – years for measuring the age and centimeters for measuring the height

Data in the long format is considered to be tidy, and in the following section we will cover the methods in Pandas to convert a dataset to this structure.

Conversion from wide to long format (tidy data)

The following are two DataFrames, with the same data but having different structures (wide and long).

Wide Format

	Biology	Chemistry	Mathematics	Physics
Andrew	90	46	95	75
Sarah	87	56	74	65
Jason	45	87	45	33

Long Format

	student_name	subject	marks
0	Andrew	Biology	90
1	Andrew	Chemistry	46
2	Andrew	Mathematics	95
3	Andrew	Physics	75
4	Sarah	Biology	87
5	Sarah	Chemistry	56
6	Sarah	Mathematics	74
7	Sarah	Physics	65
8	Jason	Biology	45
9	Jason	Chemistry	87
10	Jason	Mathematics	45
11	Jason	Physics	33

The main benefit of converting to the long format is that this format facilitates ease of data manipulation, like adding or retrieving data, as we don't need to worry about the structure of the data. Also, data retrieval is significantly faster when data is stored in a long format.

Let us understand with this an example.

First, create the following DataFrame:

CODE:

```
grades=pd.DataFrame({'Biology':[90,87,45],'Chemistry':[46,56,87],'Mathemati
cs':[95,74,45],'Physics':[75,65,33]},index=['Andrew','Sarah','Jason'])
grades
```

Output:

	Biology	Chemistry	Mathematics	Physics
Andrew	90	46	95	75
Sarah	87	56	74	65
Jason	45	87	45	33

In this DataFrame, we can see that the principles of tidy data are not followed. There are two primary variables (students and subjects) that have not been identified as columns. The values for the variable subjects, like Biology, Chemistry, Mathematics, and Physics, are observations and should not be used as columns.

Stack method (wide-to-long format conversion)

We can correct the anomalies observed in the grades DataFrame using the stack method, which takes all the column names and moves them to the index. This method returns a new DataFrame or Series, with a multilevel index.

CODE:

```
grades_stacked=grades.stack()
grades_stacked
```

Output:

```
Andrew  Biology        90
        Chemistry      46
        Mathematics    95
        Physics        75
Sarah   Biology        87
        Chemistry      56
        Mathematics    74
        Physics        65
Jason   Biology        45
        Chemistry      87
        Mathematics    45
        Physics        33
dtype: int64
```

As seen in the preceding output, the structure has been changed to a long format from a wide one.

Let us examine the data type of this stacked object.

CODE:

```
type(grades_stacked)
```

Output:

```
pandas.core.series.Series
```

As we can see, this is a Series object. We can convert this object to a DataFrame using the *reset_index* method so that the two variables - Name and Subject - can be identified as two separate columns:

CODE:

```
grades_stacked.reset_index()
```

Output:

	level_0	level_1	0
0	Andrew	Biology	90
1	Andrew	Chemistry	46
2	Andrew	Mathematics	95
3	Andrew	Physics	75
4	Sarah	Biology	87
5	Sarah	Chemistry	56
6	Sarah	Mathematics	74
7	Sarah	Physics	65
8	Jason	Biology	45
9	Jason	Chemistry	87
10	Jason	Mathematics	45
11	Jason	Physics	33

In the preceding output, we change the name of the columns using the *rename_axis* method and reset the index, as shown in the following.

CODE:

```
grades_stacked.rename_axis(['student_name','subject']).reset_
index(name='marks')
```

Output:

	student_name	subject	marks
0	Andrew	Biology	90
1	Andrew	Chemistry	46
2	Andrew	Mathematics	95
3	Andrew	Physics	75
4	Sarah	Biology	87
5	Sarah	Chemistry	56
6	Sarah	Mathematics	74
7	Sarah	Physics	65
8	Jason	Biology	45
9	Jason	Chemistry	87
10	Jason	Mathematics	45
11	Jason	Physics	33

To convert this DataFrame back to its original (wide) format, we use the unstack method, as shown in the following:

CODE:

```
grades_stacked.unstack()
```

Output:

	Biology	Chemistry	Mathematics	Physics
Andrew	90	46	95	75
Sarah	87	56	74	65
Jason	45	87	45	33

Melt method (wide-to-long format conversion)

In addition to the *stack* method, the *melt* method can also be used for converting data to the long format. The *melt* method gives more flexibility than the *stack* method by providing an option to add parameters for customizing the output.

Let us create the same DataFrame (in the wide format):

CODE:

```
grades=pd.DataFrame({'Student_Name':['Andrew','Sarah','Jason'],'Biology':
[90,87,45],'Chemistry':[46,56,87],'Mathematics':[95,74,45],'Physics':
[75,65,33]})
```

	Student_Name	Biology	Chemistry	Mathematics	Physics
0	Andrew	90	46	95	75
1	Sarah	87	56	74	65
2	Jason	45	87	45	33

Now, convert it into the long format using the *melt* method.

CODE:

```
grades.melt(id_vars='Student_Name',value_vars=['Biology','Chemistry',
'Physics','Mathematics'],var_name='Subject',value_name='Marks')
```

Output:

	Student_Name	Subject	Marks
0	Andrew	Biology	90
1	Sarah	Biology	87
2	Jason	Biology	45
3	Andrew	Chemistry	46
4	Sarah	Chemistry	56
5	Jason	Chemistry	87
6	Andrew	Physics	75
7	Sarah	Physics	65
8	Jason	Physics	33
9	Andrew	Mathematics	95
10	Sarah	Mathematics	74
11	Jason	Mathematics	45

We have used four variables in the melt method:

- id_vars: Column(s) that we don't want to reshape and preserve in its current form. If we look at the original wide format of the grades DataFrame, the Student_Name is correctly structured and can be left as it is.
- value_vars: Variable or the variables that we want to reshape into a single column. In the wide version of the grades DataFrame, there is a column for each of the four subjects. These are actually values of a single column.
- var_name: Name of the new reshaped column. We want to create a single column – "Subject", with the values "Biology", "Chemistry", "Physics", and "Mathematics".
- value_name: This is the name of the column ("Marks") containing the values corresponding to the values of the reshaped column ("Subject").

Pivot method (long-to-wide conversion)

The *pivot* method is another method for reshaping data, but unlike the *melt* and *stack* methods, this method converts the data into a wide format. In the following example, we reverse the effect of the melt operation with the pivot method.

CODE:

```
#original DataFrame
grades=pd.DataFrame({'Student_Name':['Andrew','Sarah','Jason'],'Biolo
gy':[90,87,45],'Chemistry':[46,56,87],'Mathematics':[95,74,45],'Physi
cs':[75,65,33]})
#Converting to long format with the wide method
grades_melted=grades.melt(id_vars='Student_Name',value_vars=['Biology','Che
mistry','Physics','Mathematics'],var_name='Subject',value_name='Marks')
#Converting back to wide format with the pivot method
grades_melted.pivot(index='Student_Name',columns='Subject',values='Marks')
```

Output:

Subject	Biology	Chemistry	Mathematics	Physics
Student_Name				
Andrew	90	46	95	75
Jason	45	87	45	33
Sarah	87	56	74	65

The following parameters are used with the pivot method:

- *index*: Refers to the column(s) to be used as an index. In the wide format of the grades DataFrame, we are using the Student_Name as the index.
- *columns*: Name of the column whose values are used to create a new set of columns. Each of the values of the “Subject” column forms a separate column in the wide format.

- *values*: The column used to populate the values in the DataFrame. In our grades example, the "Marks" column is used for this purpose.

Further reading:

Melt function: `https://pandas.pydata.org/pandas-docs/stable/reference/api/pandas.melt.html`

Pivot function: `https://pandas.pydata.org/pandas-docs/stable/reference/api/pandas.pivot.html`

Summary

1. Pandas, a Python library for data wrangling, offers a host of advantages, including support for a variety of input formats for data to be imported, integration with other libraries, speed, and functions for cleaning, grouping, merging, and visualizing data.

2. Series, DataFrames, and Indexes form the core of Pandas. A Series is one-dimensional and equivalent to a column, while a DataFrame is two-dimensional and equivalent to a table or spreadsheet. Series and DataFrames use indexes that are implemented using hash tables to speed up data retrieval.

3. There are various methods to create and modify Series and DataFrame objects. Python objects like lists, tuples, and dictionaries, as well as NumPy arrays, can be used to create Pandas objects. We can add and remove rows or columns, replace values, and rename columns.

4. Data retrieval in Pandas can be done by using either the position (*iloc* indexer) or the index label (*loc* indexer), or by specifying a condition (Boolean conditioning).

5. Pandas uses the *groupby* function for aggregation. Various functions can be used in conjunction with the *groupby* function to calculate aggregate measures for each group, filter the groups based on a condition, transform the values, or apply an operation on them.

6. Pandas provides support for combining, joining, and merging two or more objects. The *append* method adds an object to an existing object vertically, and the *concat* function can add objects either side by side or vertically. The *join* and *merge* methods join objects horizontally based on a common column or index values.

7. Tidy data refers to a structure where the columns correspond to variables in the dataset, the rows contain observations, and there is only one observational unit. Generally, data in the long (vertical) format is considered tidy. Functions like melt and stack convert a DataFrame into the long format. The unstack and pivot functions convert the data into the wide format.

Now that we have learned how to prepare data and make it ready for analysis, let us look at data visualization in the next chapter.

Review Exercises

Question 1

Write the function/indexer/attribute in Pandas for

- Importing data from an HTML file
- Exporting data to an Excel file
- Selecting data using the index position
- Selecting data using its label
- Replacing null values with the median
- Renaming columns
- Obtaining the number of rows and columns in a DataFrame
- Converting to the wide format using the pivot function
- Performing an inner join of two tables
- Changing the data type of a series

Question 2

The *df.describe* function returns the following summary statistical values: the count, minimum, maximum, standard deviation, percentiles, and means.

Which parameter would you add to obtain the following: the number of unique values, the value that occurs most frequently, and the frequency of this value?

Question 3

Import the "subset-covid-data.csv" in a DataFrame. Select the following data:

1. The country and continent columns
2. Set the 'country' column as the index and retrieve the population for the country "Algeria" using the at or loc indexers.
3. Select the value at the 50th row and the 3rd column using the iloc or iat indexers.
4. Retrieve the country code and population data for the last three records.
5. Select the data for the countries where the population is greater than 2.5 million, and the number of cases is greater than 3000.

Question 4

Import the data for the file "subset-covid-data.csv" in a DataFrame and write code statements for the following:

1. Deleting the "country_code" column
2. Inserting this column back at the same position that it was present earlier
3. Deleting the first three rows
4. Adding these rows back (at their original position)

Question 5

Create the following DataFrames:

```
DataFrame name: orders_df
```

	order_id	item
0	1	pens
1	2	shirts
2	3	coffee

```
DataFrame name: orders1_df
```

	order_id	item
0	4	crayons
1	5	tea
2	6	fruits

```
DataFrame name:customers_df
```

	order id	customer_name
0	1	anne
1	2	ben
2	3	carlos

Which function or method would you use to:

1. Combine the details of the first two DataFramesorders_df and orders1_df?
2. Create a DataFrame to show the customers and the items they ordered?

3. Make the order_id column as the index for orders_df and customers_df? Which method would you now use to combine these two objects to show which orders were placed by customers?

Question 6

The following DataFrame records the weight fluctuations of four people:

	Anna	Ben	Carole	Dave
0	51.0	70.0	64.0	81.0
1	52.0	70.5	64.2	81.3
2	51.4	69.1	66.8	80.5
3	52.8	69.8	66.0	80.9
4	50.5	70.5	63.4	81.4

1. Create the preceding DataFrame.
2. Convert this DataFrame into a tidy format.
3. Determine who among these four people had the least fluctuation in weight.
4. For people whose average weight is less than 65 kgs, convert their weight (on all four days) into pounds and display this data.

Question 7

The object data type is used for columns that contain the following data:

1. Strings
2. Numbers (int and float)
3. Booleans
4. A combination of all the preceding data types

Question 8

A column can be accessed using the dot notation (a.column_name) as well as the indexing operator (a[column_name]). Which is the preferred notation, and why?

Question 9

Which method is used to obtain the metadata in a DataFrame?

1. Describe method
2. Value_counts method
3. Info method
4. None of the above

Question 10 (fill in the blanks)

- The default join type for the join method is ____ and the parameter for adding a join type is ____
- The default join type for the merge method is ____ and the parameter for adding a join type is ____
- The default join type for the concat function is ____ and the parameter for adding a join type is ____
- The function in Pandas that created an object representing a date/time value and is equivalent to the following Python functions: datetime.date, datetime.time, or datetime.datetime is ____
- The function in Pandas equivalent to the datetime.timedelta function in Python for representing a duration of time is ____

Answers

Question 1

- Importing data from an HTML file:

  CODE:

  ```
  pd.read_html(url)
  ```

- Exporting data to an Excel file:

  CODE:

  ```
  df.to_excel(name_of_file)
  ```

- Selecting data using the index position:

  CODE:

  ```
  df.iloc[index position]
  ```

- Selecting data using its label:

  CODE:

  ```
  df.loc[column name or index label]
  ```

- Replacing null values in a DataFrame with the median:

  CODE:

  ```
  df.fillna(df.median())
  ```

- Renaming columns:

  ```
  df.rename(columns={'previous_name:'new_name'})
  ```

- Obtaining the number of rows and columns in a DataFrame:

  CODE:

  ```
  df.shape
  ```

- Converting to the wide format using the *pivot* function:

  CODE:

  ```
  df.pivot(index=column_name1,
  columns=column_name2, values=column_name3)
  ```

- Performing an inner join of two DataFrame objects

  CODE:

  ```
  df1.merge(df2)
  ```

  The default join type is *inner* for the *merge* method; hence we need not explicitly mention it.

- Changing the data type of a Series object/column:

  CODE:

  ```
  series_name.astype(new data type)
  ```

Question 2

We can obtain all three values using the include='all' parameter, as shown in the following:

CODE:

```
df.describe(include='all')
```

Question 3

CODE:

```
df=pd.read_csv('subset-covid-data.csv')
```

1.Retrieving the country and continent columns

CODE:

```
df[['country','continent']]
```

2.Set the 'country' column as the index and retrieve the population for the country "Algeria" using the at or loc indexers.

CODE:

```
df.set_index('country',inplace=True)
```

Retrieve the population using this statement:

CODE:

```
df.at['Algeria','population']
```

or

```
df.loc['Algeria','population']
```

3.Select the value at the 50th row and the 3rd column using the iloc or iat indexers.

```
df.iat[49,2]
```

Or

```
df.iloc[49,2]
```

4.Retrieve the country code and population data for the last three records:

CODE:

```
df.iloc[203:,-1:-3:-1]
```

Or

CODE:

```
df.iloc[203:,7:]
```

5.Select the data for the countries where the population is greater than 2.5 million, and the number of cases is greater than 3000.

CODE:

```
df[(df['cases']>=3000) & (df['population']>=25000000)]
```

Question 4 (addition and deletion of rows and columns)

1. Delete the "country_code" column:

   CODE:

   ```
   x=df.pop('country_code')
   ```

2. Insert this column back at the same position that it was present earlier:

   CODE:

   ```
   df.insert(8,'country_code',x)
   ```

3. Delete the first three rows:

   CODE:

   ```
   df.drop([0,1,2],inplace=True)
   ```

4. Add these rows back to the DataFrame at their original positions:

   CODE:

   ```
   x=df.iloc[0:3]
   pd.concat([x,df])
   ```

Question 5

Create the following DataFrames:

CODE:

```
orders_df=pd.DataFrame({'order_id':[1,2,3],'item':['pens','shirts',
'coffee']})
orders1_df=pd.DataFrame({'order_id':[4,5,6],'item':['crayons','tea',
'fruits']})
customers_df=pd.DataFrame({'order id':[1,2,3],'customer_name':['anne',
'ben','carlos']})
```

Functions for combining objects

- Combine the details of the two DataFrames orders_df and orders1_df:

  CODE:

  ```
  pd.concat((orders_df,orders1_df))
  ```

- Create a combined DataFrame to show the customers and the items they ordered:

  CODE:

```
pd.merge(orders_df,customers_df,left_on='order_id',right
_on='order id')
```

- Make the order_id/order id column as the index for the "orders_df" and "customers_df" DataFrames:

  CODE:

```
orders_df.set_index('order_id',inplace=True)
customers_df.set_index('order id',inplace=True)
```

- Method to combine these two objects to show which orders were placed by customers:

  CODE:

```
orders_df.join(customers_df)
```

Question 6 (tidy data and aggregation)

1. Create this DataFrame.

   CODE:

```
df=pd.DataFrame({'Anna':[51,52,51.4,52.8,50.5],'Ben':
[70,70.5,69.1,69.8,70.5],'Carole':[64,64.2,66.8,66,63.4],
'Dave':[81,81.3,80.5,80.9,81.4]})
```

2. Convert this into a tidy format.

   We use the *melt* method to convert the DataFrame to a long format and then rename the columns.

   CODE:

```
df_melted=df.melt()
df_melted.columns=['name','weight']
```

3. In order to find out who among these four people had the least fluctuation in weight with respect to their mean weight, we need to first aggregate the data.

   We use the *groupby* function to group the data according to the name of the person and use the standard deviation (*np.std*) aggregation function.

   CODE:

   ```
   df_melted.groupby('name').agg({'weight':np.std})
   ```

   Output:

	weight
name	
Anna	0.893308
Ben	0.580517
Carole	1.446375
Dave	0.356371

   Dave had the least standard deviation as seen from the preceding output, and therefore, the least fluctuation in his weight.

4. For people whose average weight is less than 65 kgs, we now need to convert their weight to pounds and display their weight for all the four days.

   We use the filter method to filter out the groups where the average weight is greater than 65 kgs using the filter method and then apply the transform method to convert the weight into pounds.

   CODE:

   ```
   df_melted.groupby('name').filter(lambda x:x.mean()>65)
   ['weight'].transform(lambda x:float(x)*2.204)
   ```

Question 7

Options 1 and 4

The object data type in Pandas is used for strings or columns that contain mixed data like numbers, strings, and so on.

Question 8

The indexing operator [] is preferred since it does not clash with built-in methods and works with column names that have spaces and special characters. The dot notation does not work when you have to retrieve a column that contains spaces.

Question 9

Option 3 - info method.

The info method is used to obtain the metadata for a Series or DataFrame object. It gives us the name of the columns, their data types, and number of non-null values and the memory usage.

Question 10 (fill in the blanks)

The default join type for the join method is a *"left join"* and the parameter for adding a join type is the *"how"* parameter

The default join type for the merge method is an *"inner join"* and the parameter for adding a join type is the *"how"* parameter

The default join type for the concat function is an *"outer join"* and the parameter for adding a join type is the *"join"* parameter

The function in Pandas that created an object representing a date/time value and is equivalent to the following Python functions: *datetime.date*, *datetime.time*, or *datetime.datetime* is *pd.Timestamp*

The function in Pandas equivalent to the *datetime.timedelta* function in Python for representing a duration of time is *pd.Timedelta*

CHAPTER 7

Data Visualization with Python Libraries

In the last chapter, we read about Pandas, the library with various functions for preparing data in order to make it ready for analysis and visualization. Visualization is a means to understand patterns in your data, identify outliers and other points of interest, and present our findings to an outside audience, without having to sift through data manually. Visualization also helps us to glean information from raw data and gain insights that would otherwise be difficult to draw.

After going through this chapter, you will be able to understand the commonly used plots, comprehend the object-oriented and stateful approaches in Matplotlib and apply these approaches for visualization, learn how to use Pandas for plotting, and understand how to create graphs with Seaborn.

Technical requirements

In your Jupyter notebook, type the following to import the following libraries.

CODE:

```
import pandas as pd
import seaborn as sns
import matplotlib.pyplot as plt
```

Here, *plt* is a shorthand name or an alias for the *pyplot* module of Matplotlib that we use for plotting, *sns* is an alias for the Seaborn library, and *pd* is an alias for Pandas.

G. Rajagopalan, *A Python Data Analyst's Toolkit*, https://doi.org/10.1007/978-1-4842-6399-0_7

In case these libraries are not installed, go to the Anaconda Prompt and install them as follows:

```
pip install matplotlib
pip install seaborn
pip install pandas
```

External files

We use the *Titanic* dataset in this chapter to demonstrate the various plots.

Please download the dataset using the following link: `https://github.com/DataRepo2019/Data-files/blob/master/titanic.csv`

You can also download this dataset using the following steps:

- Click the following link: `https://github.com/DataRepo2019/Data-files`
- Click: Code ➤ Download ZIP

- From the downloaded zip folder, open the "titanic.csv" file

Commonly used plots

Some of the basic plots that are widely used in exploratory or descriptive data analysis include bar plots, pie charts, histograms, scatter plots, box plots, and heat maps; these are explained in Table 7-1.

Table 7-1. *Commonly Used Plots to Visualize Data in Descriptive Data Analysis*

Type of Chart or Plot	Description	Shape
Bar chart	A bar chart enables visualization of categorical data, with the width or height of the bar representing the value for each category. The bars can be shown either vertically or horizontally.	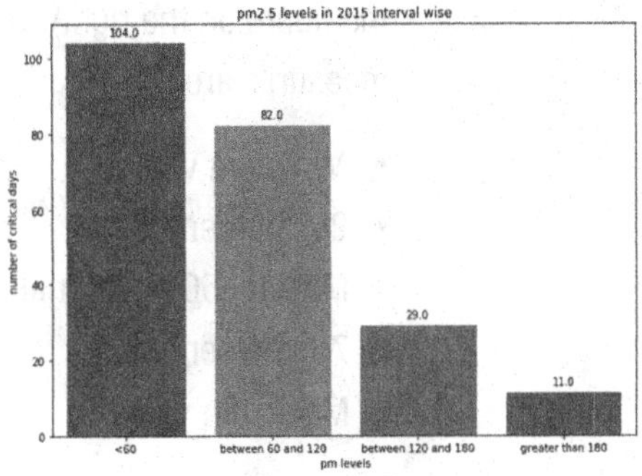
Histogram	A histogram is used to visualize the distribution of a continuous variable. It divides the range of the continuous variable into intervals and shows where most of the values lie.	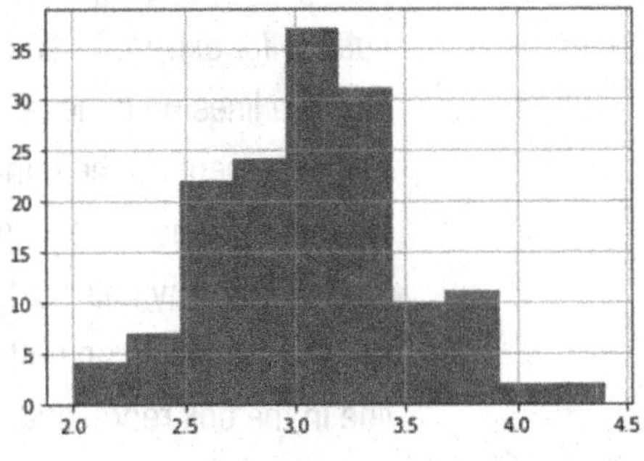

(continued)

Table 7-1. *(continued)*

Type of Chart or Plot	Description	Shape
Box plots	Box plots help with visually depicting the statistical characteristics of the data. A box plot provides a five-point summary with each line in the figure representing a statistical measure of the data being plotted (refer to the figure on the right). These five measures are • Minimum value • 25th percentile • Median (50th percentile) • 75th percentile • Maximum value The small circles/dots that you see in the figure on the right represent the outliers (or extreme values). The two lines on either side of the box, representing the minimum and maximum values, are also called "whiskers". Any point outside these whiskers is called an outlier. The middle line in the box represents the median. A box plot is generally used for continuous (ratio/interval) variables, though it can be used for some categorical variables like ordinal variables as well.	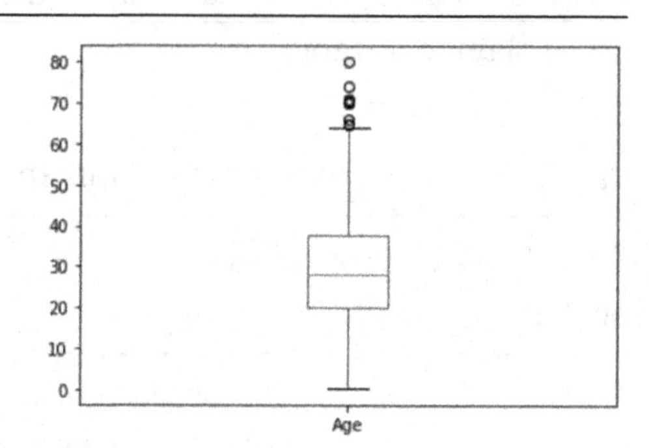

(continued)

Table 7-1. *(continued)*

Type of Chart or Plot	Description	Shape
Pie charts	A pie chart shows the distinct values of a variable as sectors within a circle. Pie charts are used with categorical variables.	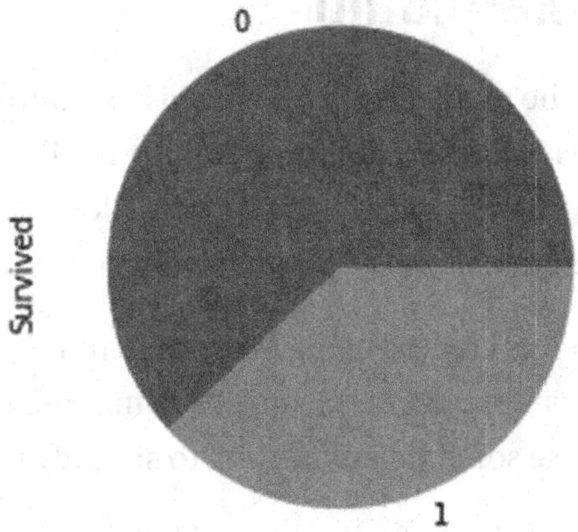
Scatter plots	A scatter plot displays the values of two continuous variables as points on the x and y axes and helps us visualize if the two variables are correlated or not.	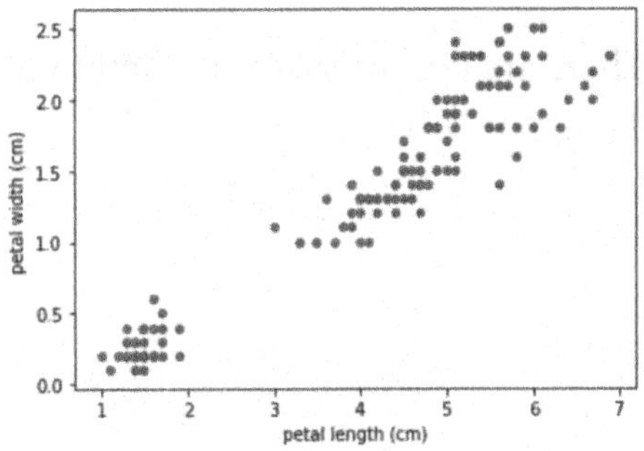
Heat maps	A heat map shows the correlation between multiple variables using a color-coded matrix, where the color saturation represents the strength of the correlation between the variables. A heat map can aid in the visualization of multiple variables at once.	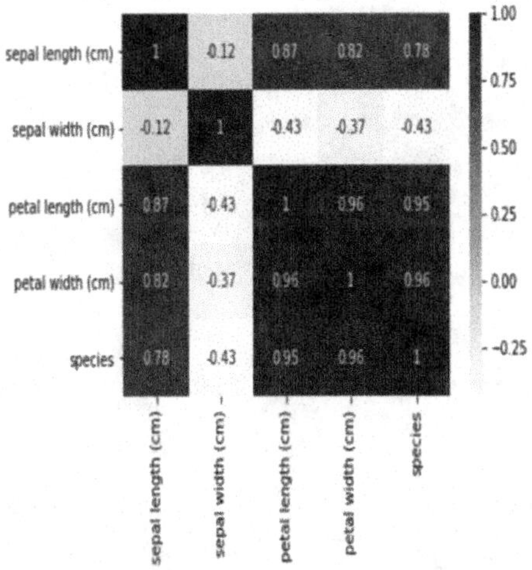

Let us now have a look at some of the Python libraries that are used for visualization, starting with Matplotlib.

Matplotlib

The main library for data visualization in Python is Matplotlib. Matplotlib has many visualization features that are similar to Matlab (a computational environment cum programming language with plotting tools). Matplotlib is mainly used for plotting two-dimensional graphs, with limited support for creating three-dimensional graphs.

Plots created using Matplotlib require more lines of code and customization of the parameters of the plot, as compared to other libraries like Seaborn and Pandas (which use some default setting to simplify the writing of code to create plots).

Matplotlib forms the backbone of most of the visualizations that are performed using Python.

There are two interfaces in Matplotlib, stateful and object-oriented, that are described in Table 7-2.

Table 7-2. *Interfaces in Matplotlib*

Stateful interface	Object-oriented interface
This interface uses the *pyplot* class, which is based on Matlab. A single object of this class is created, which is used for all plotting purposes.	In this interface, we use different objects for different elements of the plot. The two main objects that are used in this interface for plotting are • The *figure* object, which acts as the container for other objects. • The *axes* object, which is the actual plotting area containing the x axis, y axis, points, lines, legends, and labels. Note that the axes here does not refer to the x and y axes but the entire subplot.

Code Example (visualization using stateful interface):
`import matplotlib.pyplot as plt`
`%matplotlib inline`
`x=[1,2,3,4]`
`y=[3,6,9,12]`
`plt.plot(x,y) # The plot function plots a line between the x and y coordinates`
`plt.xlim(0,5) # Sets limits for the x axis`
`plt.ylim(0,15) # Sets limits for the y axis`
	Code Example (visualization using object-oriented interface):
	`import matplotlib.pyplot as plt`
	`%matplotlib inline`
	`x=[1,2,3,4]`
	`y=[3,6,9,12]`
	`fig,ax=plt.subplots(figsize=(10,5)) #The subplots method creates a tuple returning a figure and one or more axes.`
	`ax.plot(x,y) #Creating a plot with an axes object`

(*continued*)

Table 7-2. *(continued)*

Stateful interface	Object-oriented interface
plt.xlabel('X axis') #labels the x axis plt.ylabel('Y axis') #labels the y axis plt.title('Basic plot') #Gives a title plt.suptitle('Figure title',size=20,y=1.02) # Gives a title to the overall figure	
Customization (with the stateful interface): In this interface, all changes are made using the current state of the pyplot object that points to the figure or axes, as shown in the following.	Customization (with the object-oriented interface): In this interface, since we have different objects for the figure and each of the subplots or axes, these objects are customized and labeled individually, as shown in the following.
Code Example: #This code makes changes to the graph created using the plt object ax=plt.gca() # current axes ax.set_ylim(0,12) #use it to set the y axis limits fig=plt.gcf() #current figure fig.set_size_inches(4,4) #use it to set the figure size	Code Example: #this code makes changes to the graph created using the preceding object-oriented interface ax.set_xlim(0,5) # Sets limit for the x axis ax.set_ylim(0,15) # Sets limit for the y axis ax.set_xlabel('X axis') #Labels x axis ax.set_ylabel('Y axis') #Labels y axis ax.set_title('Basic plot') # Gives a title to the graph plotted fig.suptitle('Figure title',size=20,y=1.03) #Gives a title to the overall figure

Further reading: See more about these two different interfaces: `https://matplotlib.org/3.1.1/tutorials/introductory/lifecycle.html`

Approach for plotting using Matplotlib

The object-oriented approach is the recommended approach for plotting in Matplotlib because of the ability to control and customize each of the individual objects or plots. The following steps use the object-oriented methodology for plotting.

1. **Create a figure (the outer container) and set its dimensions**:

   The *plt.figure* function creates a figure along with setting its dimensions (width and height), as shown in the following.

   CODE:

   ```
   fig=plt.figure(figsize=(10,5))
   ```

2. **Determine the number of subplots and assign positions for each of the subplots in the figure**:

   In the following example, we are creating two subplots and placing them vertically. Hence, we divide the figure into two rows and one column with one subplot in each section.

   The *fig.add_subplot* function creates an axes object or subplot and assigns a position to each subplot. The argument –211 (for the *add_subplot* function that creates the first axes object - "ax1") means that we are giving it the first position in the figure with two rows and one column.

   The argument -212 (for the *add_subplot* function that creates the second axes object - "ax2") means that we are giving the second position in the figure with two rows and one column. Note that the first digit indicates the number of rows, the second digit indicates the number of columns, and the last digit indicates the position of the subplot or axes.

CODE:

```
ax1=fig.add_subplot(211)
ax2=fig.add_subplot(212)
```

3. **Plot and label each subplot**:

   After the positions are assigned to each subplot, we move on to generating the individual subplots. We are creating one histogram (using the *hist* function) and one bar plot (using the *bar* function). The x and y axes are labeled using the *set_xlabel* and *set_ylabel* functions.

   CODE:

```
labelling the x axis
ax1.set_xlabel("Age")
#labelling the yaxis
ax1.set_ylabel("Frequency")
#plotting a histogram using the hist function
ax1.hist(df['Age'])
#labelling the X axis
ax2.set_xlabel("Category")
#labelling the Y axis
ax2.set_ylabel("Numbers")
#setting the x and y lists for plotting the bar chart
x=['Males','Females']
y=[577,314]
#using the bar function to plot a bar graph
ax2.bar(x,y)
```

Output:

Note that the top half of Figure 7-1 is occupied by the first axes object (histogram), and the bottom half of the figure contains the second subplot (bar plot).

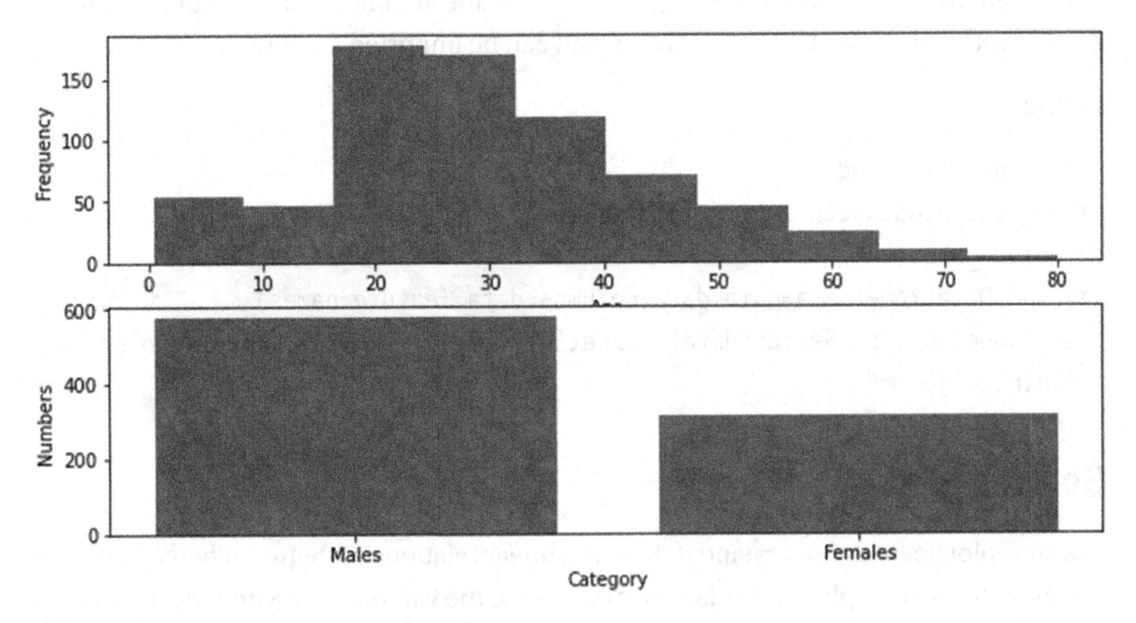

Figure 7-1. *Subplots within a figure*

Plotting using Pandas

The Pandas library uses the Matplotlib library behind the scenes for visualizations, but plotting graphs using Pandas functions is much more intuitive and user-friendly. Pandas requires data to be in the wide or aggregated format.

The *plot* function (based on the Matplotlib plot function) used in Pandas allows us to create a wide variety of plots simply by customizing the value of the *kind* parameter, which specifies the type of plot. This is also an example of polymorphism in object-oriented programming (one of the principles of OOPS, which we studied in Chapter 2), where we are using the same method for doing different things. The *kind* parameter in the *plot* method changes with the kind of graph you want to plot.

Let us learn how to create plots in Pandas using the *Iris* dataset.

The *Iris* dataset contains samples from various species of the iris plant. Each sample contains five attributes: sepal length, sepal width, petal length, petal width, and species (*Iris setosa, Iris versicolor,* and *Iris virginica*). There are 50 samples of each species. The *Iris* dataset is inbuilt in the *sklearn* library and can be imported as follows:

CODE:

```
import pandas as pd
from sklearn.datasets import load_iris
data=load_iris()
iris=pd.DataFrame(data=data.data,columns=data.feature_names)
iris['species']=pd.Series(data['target']).map({0:'setosa',1:'versicolor',2:
'virginica'})
```

Scatter plot

A scatter plot helps us understand if there is a linear relationship between two variables. To generate a scatter plot in Pandas, we need to use the value *scatter* with the parameter *kind* and mention the columns (specified by the parameters "x" and "y") to be used for plotting in the argument list of the *plot* function. The graph in Figure 7-2 suggests that the two variables ("petal length" and "petal width") are linearly correlated.

CODE:

```
iris.plot(kind='scatter',x='petal length (cm)',y='petal width (cm)')
```

Output :

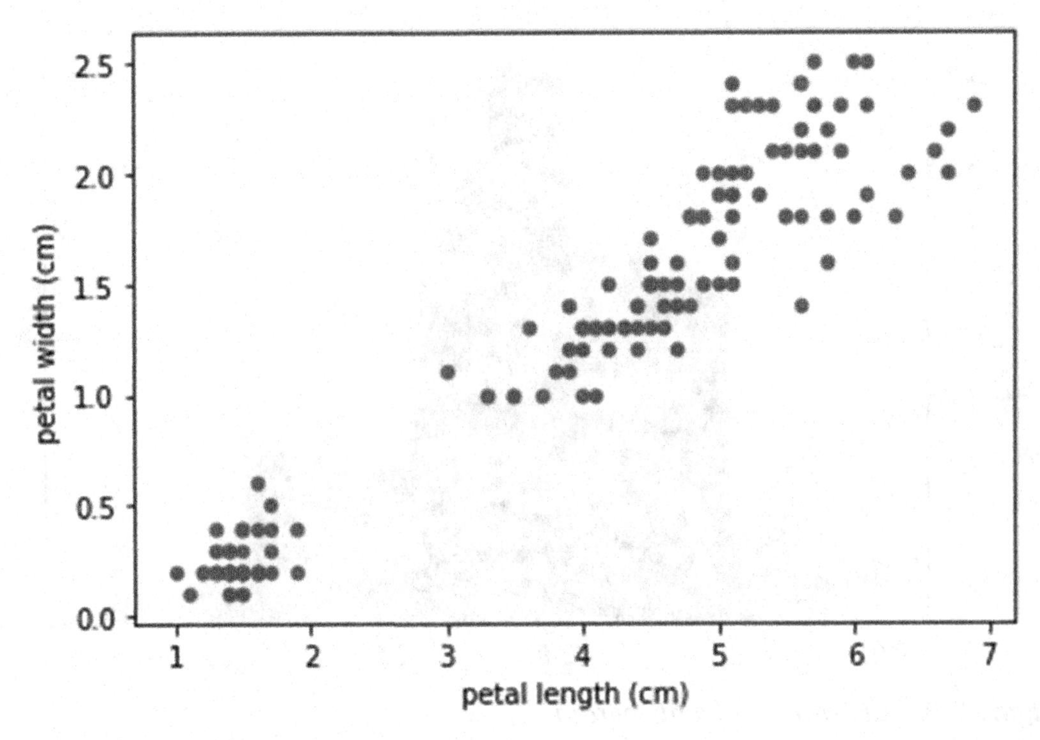

Figure 7-2. *Scatter plot in Pandas*

Histogram

A histogram is used to visualize the frequency distribution with various bars representing the frequencies across various intervals (Figure 7-3). The value 'hist' is used with the *kind* parameter in the *plot* function to create histograms.

CODE:

```
iris['sepal width (cm)'].plot(kind='hist')
```

Output:

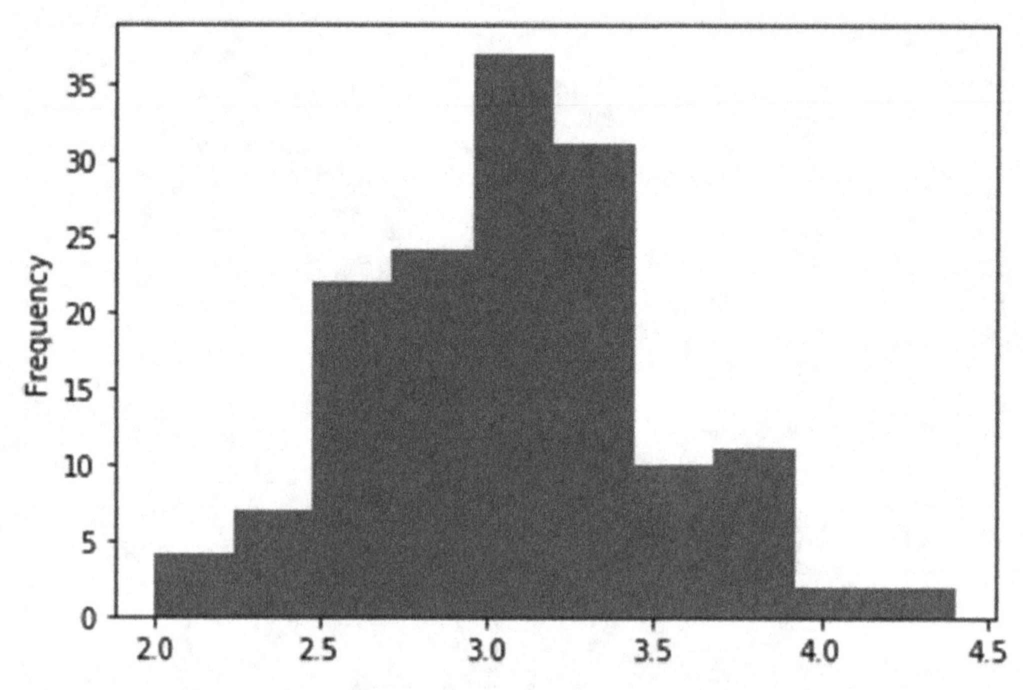

Figure 7-3. *An example of a histogram*

As we can see from this histogram, the "sepal width" variable is normally distributed approximately.

Pie charts

A pie chart shows different values that form a variable as sectors in a circle (Figure 7-4). Note that Pandas requires the *value_counts* function to calculate the number of values in each category as aggregation is not performed on its own when plotting is done in Pandas (we will later see that aggregation is taken care of if plotting is done using the Seaborn library). We need to use the value "pie" with the *kind* parameter to create pie charts.

CODE:

```
iris['species'].value_counts().plot(kind='pie')
```

Output:

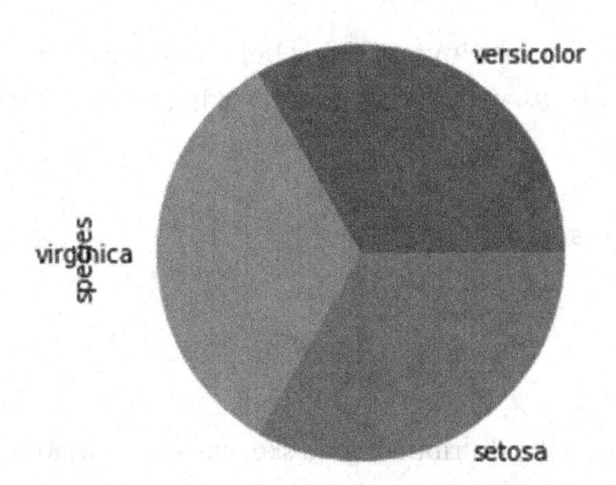

Figure 7-4. *Creating a pie chart with Pandas*

We see that the three species ("virginica", "setosa", and "versicolor") form equal parts of a circle, that is, they have the same number of values.

The Pandas *plot* method is very intuitive and easy to use. By merely changing the value of the *kind* parameter, we can plot a variety of graphs.

Further reading: See more about the kinds of plots that can be used in Pandas:

`https://pandas.pydata.org/pandas-docs/stable/user_guide/visualization.html#other-plots`

Seaborn library

Seaborn is another Python-based data visualization library. Seaborn changes the default properties of Matplotlib to adjust the color palettes and perform aggregation automatically on columns. The default settings make it easier to write the code needed for creating various plots.

Seaborn offers the ability to customize these plots as well, but the customization options are less as compared to Matplotlib.

Seaborn enables the visualization of data in more than two dimensions. It also requires data to be in the long (tidy) format, which is in contrast to Pandas, which needs data to be in a wide form.

Let us see how to plot graphs using Seaborn with the *Titanic* dataset.

We use the functions in Seaborn to create different plots for visualizing different variables in this dataset.

The Seaborn library needs to be imported first before its functions can be used. The alias for the Seaborn library is *sns*, which is used for invoking the plotting functions.

CODE:

```
import seaborn as sns
titanic=pd.read_csv('titanic.csv')
```

Box plots

A box plot gives an idea of the distribution and skewness of a variable, based on statistical parameters, and indicates the presence of outliers (denoted by circles or dots), as shown in Figure 7-5. The *boxplot* function in Seaborn can be used to create box plots. The column name of the feature to be visualized is passed as an argument to this function.

CODE:

```
sns.boxplot(titanic['Age'])
```

Output:

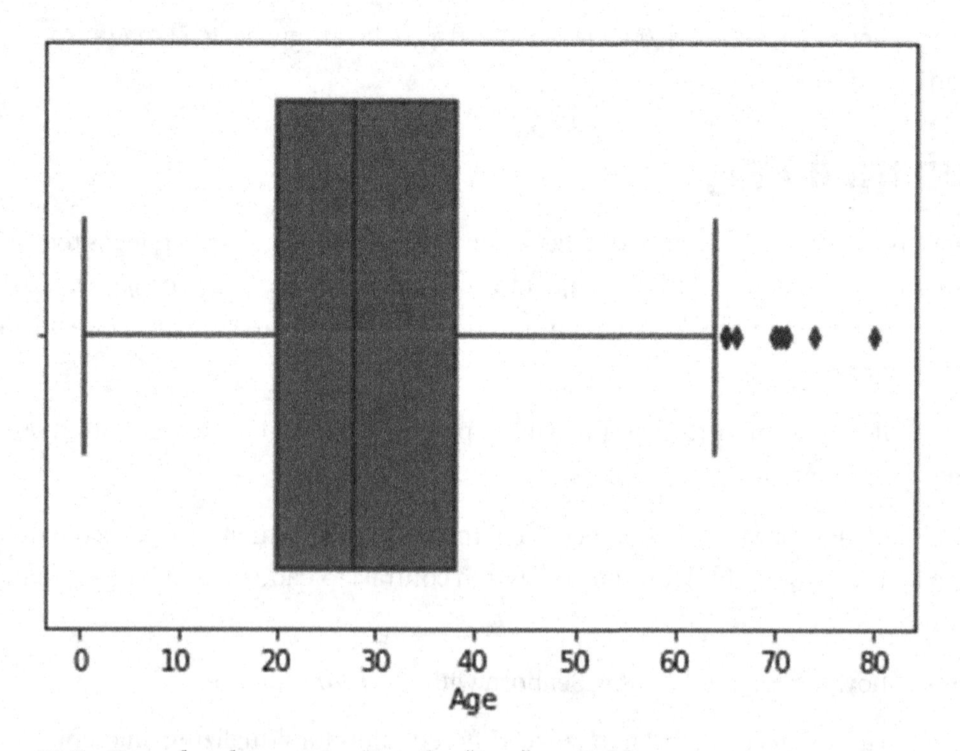

Figure 7-5. *Box plot demonstrating the "Age" variable*

Adding arguments to any Seaborn plotting function

There are two methods we can use when we pass arguments to any function used in Seaborn:

- We can either use the full column name (that includes the name of the DataFrame), skipping the *data* parameter.

  CODE:

  ```
  sns.boxplot(titanic['Age'])
  ```

- Or, mention the column names as strings and use the *data* parameter to specify the name of the DataFrame.

  CODE:

  ```
  sns.boxplot(x='Age',data=titanic)
  ```

Kernel density estimate

The kernel density estimate is a plot for visualizing the probability distribution of a continuous variable, as shown in Figure 7-6. The *kdeplot* function in Seaborn is used for plotting a kernel density estimate.

CODE:

```
sns.kdeplot(titanic['Age'])
```

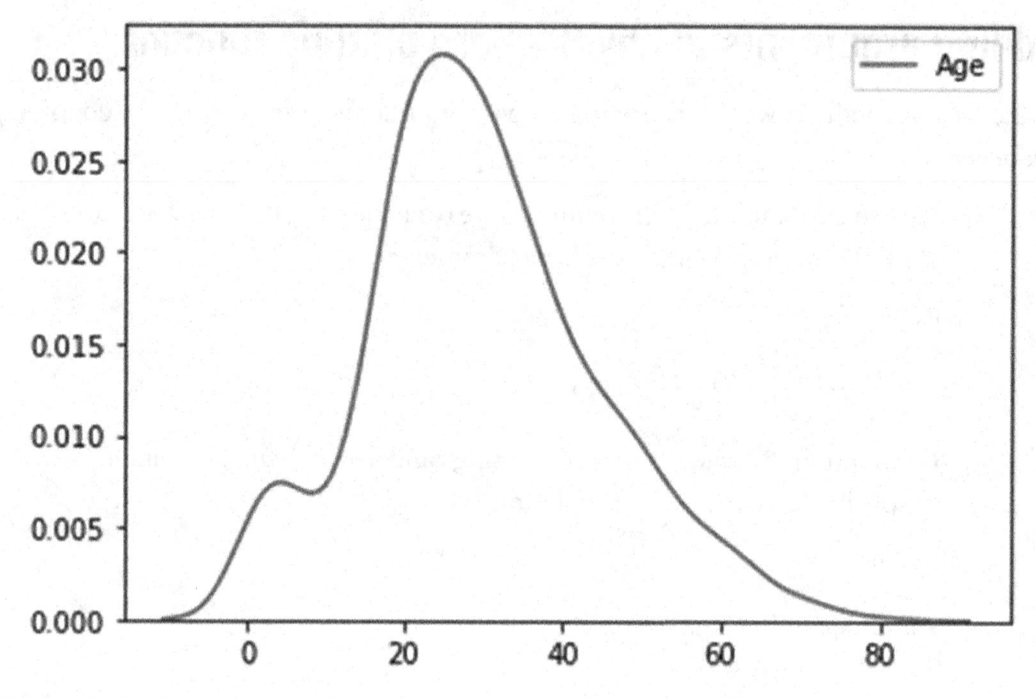

Figure 7-6. *An example of a kernel density estimate (KDE) plot*

Further reading: `https://seaborn.pydata.org/generated/seaborn.kdeplot.html`

Violin plot

A violin plot merges the box plot with the kernel density plot, with the shape of the violin representing the frequency distribution, as shown in Figure 7-7. We use the *violinplot* function in Seaborn for generating violin plots.

CODE:

```
sns.violinplot(x='Pclass',y='Age',data=titanic)
```

Output:

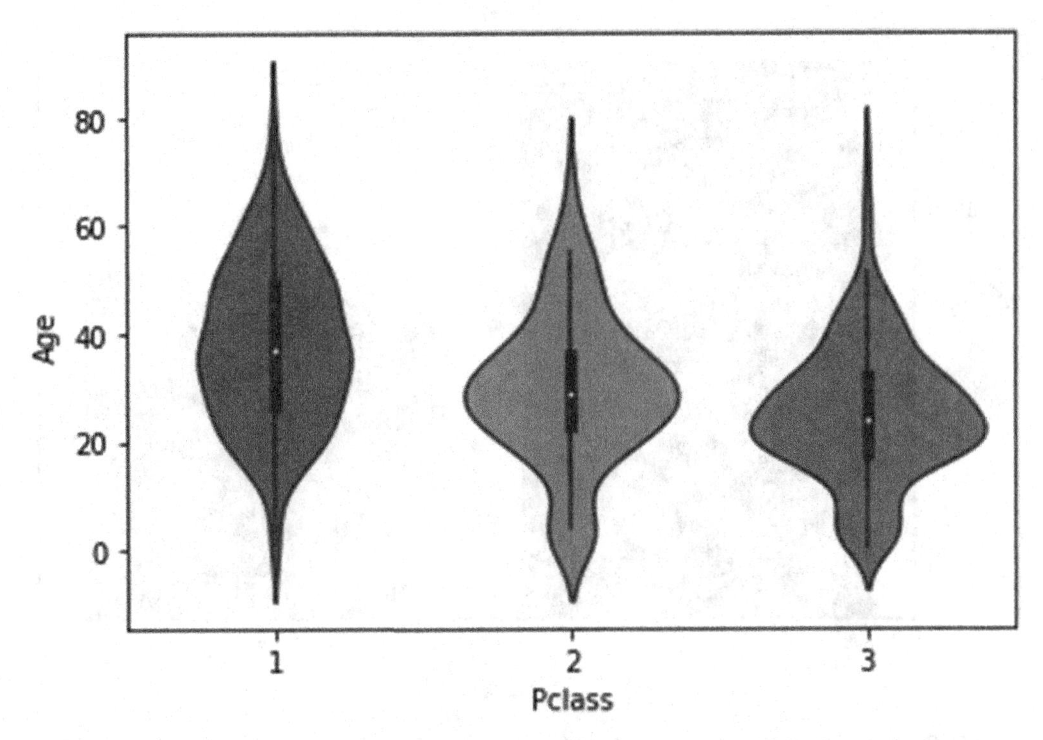

Figure 7-7. *An example of a violin plot in Seaborn*

Further reading: `https://seaborn.pydata.org/generated/seaborn.violinplot.html`

Count plots

Count plots are used to plot categorical variables, with the length of the bars representing the number of observations for each unique value of the variable. In Figure 7-8, the two bars are showing the number of passengers who did not survive (corresponding to a value of 0 for the "Survived" variable) and the number of passengers who survived (corresponding to a value of 1 for the "Survived" variable). The *countplot* function in Seaborn can be used for generating count plots.

CODE:

```
sns.countplot(titanic['Survived'])
```

Output:

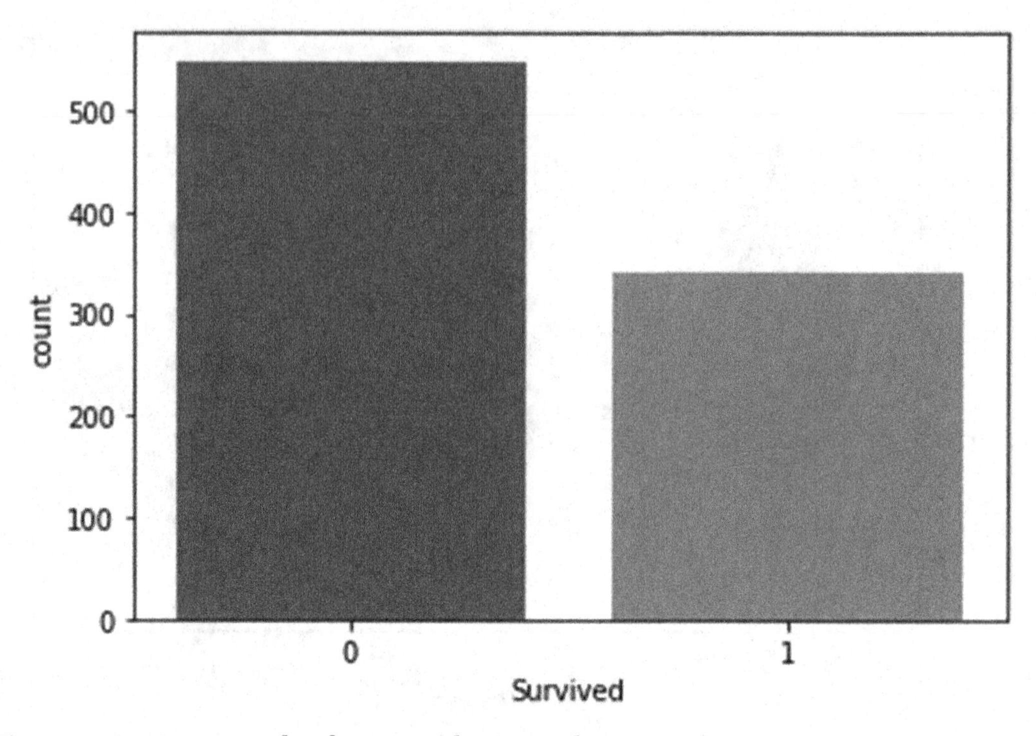

Figure 7-8. *An example of a countplot in Seaborn*

Further reading: `https://seaborn.pydata.org/generated/seaborn.countplot.html`

Heatmap

A heatmap is a graphical representation of a correlation matrix, representing the correlation between different variables in a dataset, as shown in Figure 7-9. The intensity of the color represents the strength of the correlation, and the values represent the degree of correlation (a value closer to one represents two strongly correlated variables). Note that the values along the diagonal are all one since they represent the correlation of the variable with itself.

The *heatmap* function in Seaborn creates the heat map. The parameter *annot* (with the value "True") enables the display of values representing the degree of correlation, and the *cmap* parameter can be used to change the default color palette. The *corr* method creates a DataFrame containing the degree of correlation between various pairs of variables. The labels of the heatmap are populated from the index and column values in the correlation DataFrame (titanic.corr in this example).

CODE:

```
sns.heatmap(titanic.corr(),annot=True,cmap='YlGnBu')
```

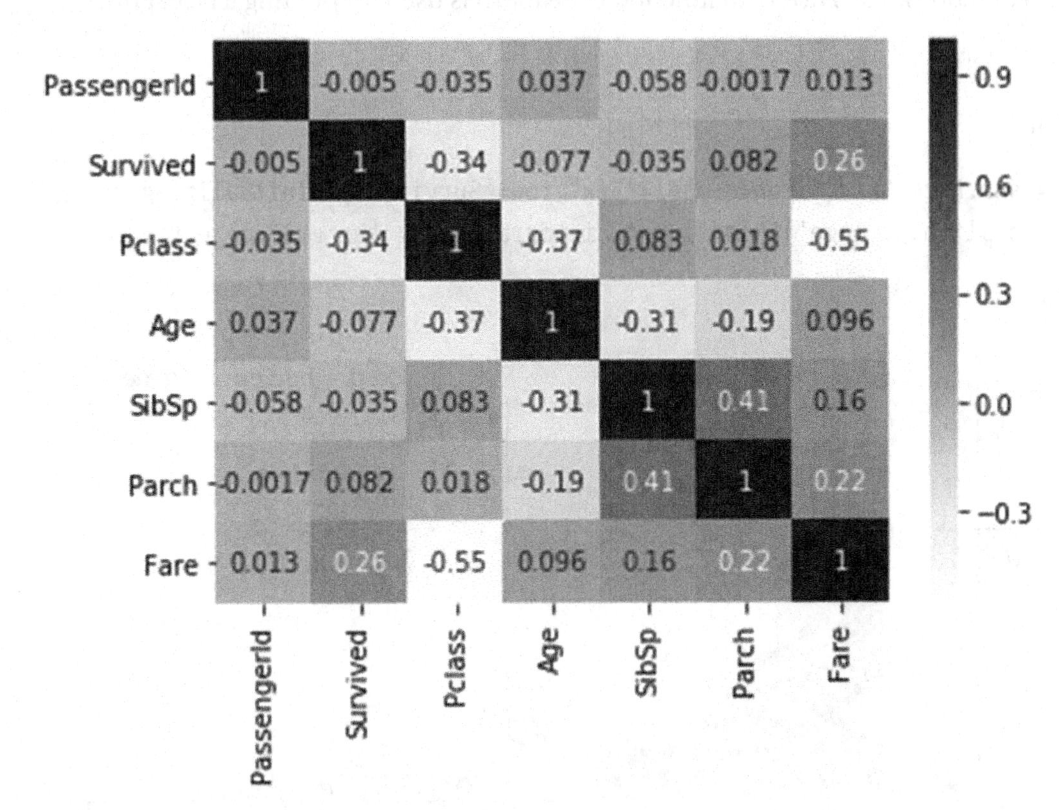

Figure 7-9. *An example of a heatmap in Seaborn*

Further reading:

See more about the "heatmap" function and its parameters:

`https://seaborn.pydata.org/generated/seaborn.heatmap.html`

See more about customizing color palettes and color maps: `https://seaborn.pydata.org/tutorial/color_palettes.html`

Facet grid

A facet grid represents the distribution of a single parameter or the relationship between parameters across a grid containing a *row*, *column*, or *hue* parameter, as shown in Figure 7-10. In the first step, we create a grid object (the *row*, *col*, and *hue* parameters are optional), and in the second step, we use this grid object to plot a graph of our choice (the name of the plot and the variables to be plotted are supplied as arguments to the map function). The *FacetGrid* function in Seaborn is used for plotting a facet grid.

Example:

CODE:

```
g = sns.FacetGrid(titanic, col="Sex",row='Survived') #Initializing the grid
g.map(plt.hist,'Age')#Plotting a histogram using the grid object
```

Output:

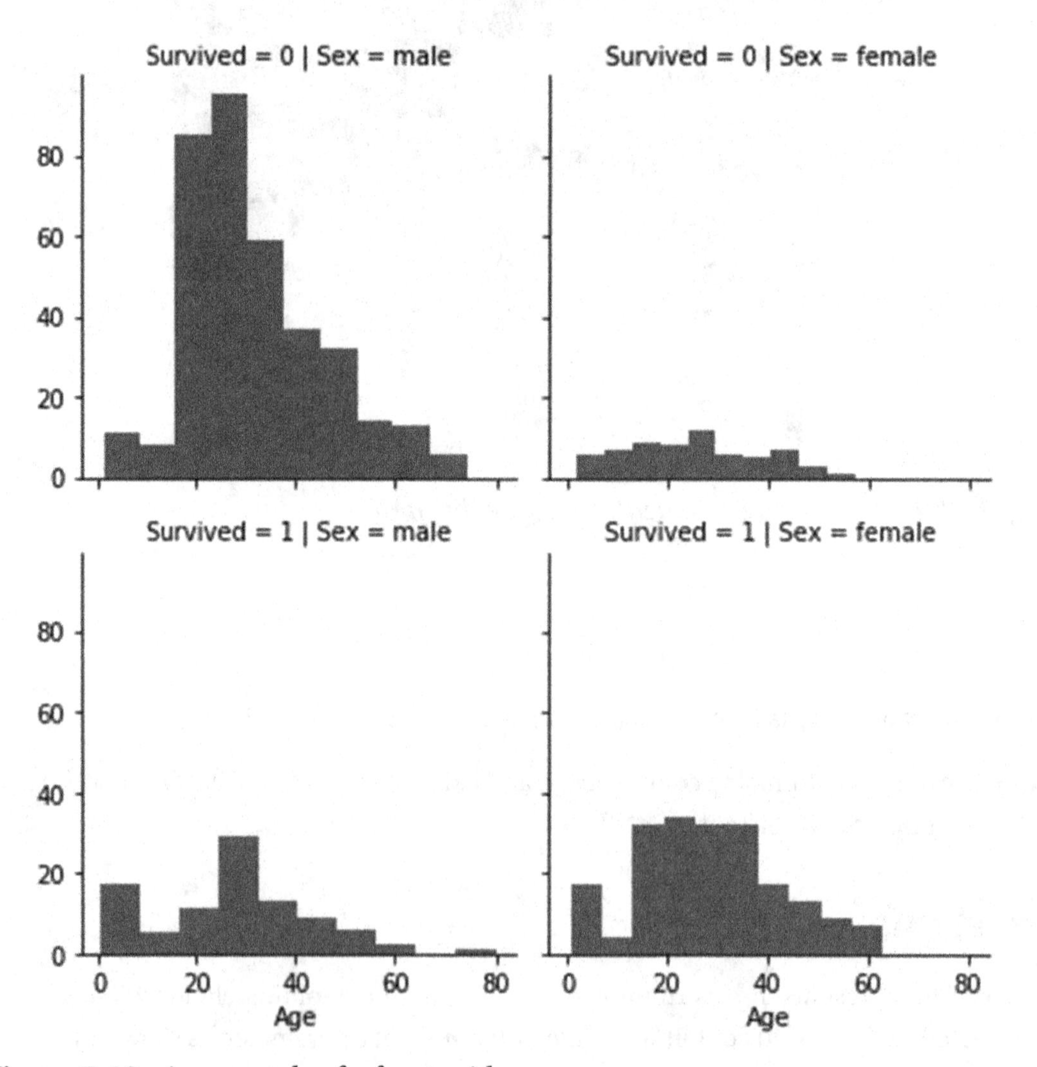

Figure 7-10. *An example of a facet grid*

Regplot

This plot uses the linear regression model to plot a regression line between the data points of two continuous variables, as shown in Figure 7-11. The Seaborn function *regplot* is used for creating this plot.

CODE:

```
sns.regplot(x='Age',y='Fare',data=titanic)
```

Output:

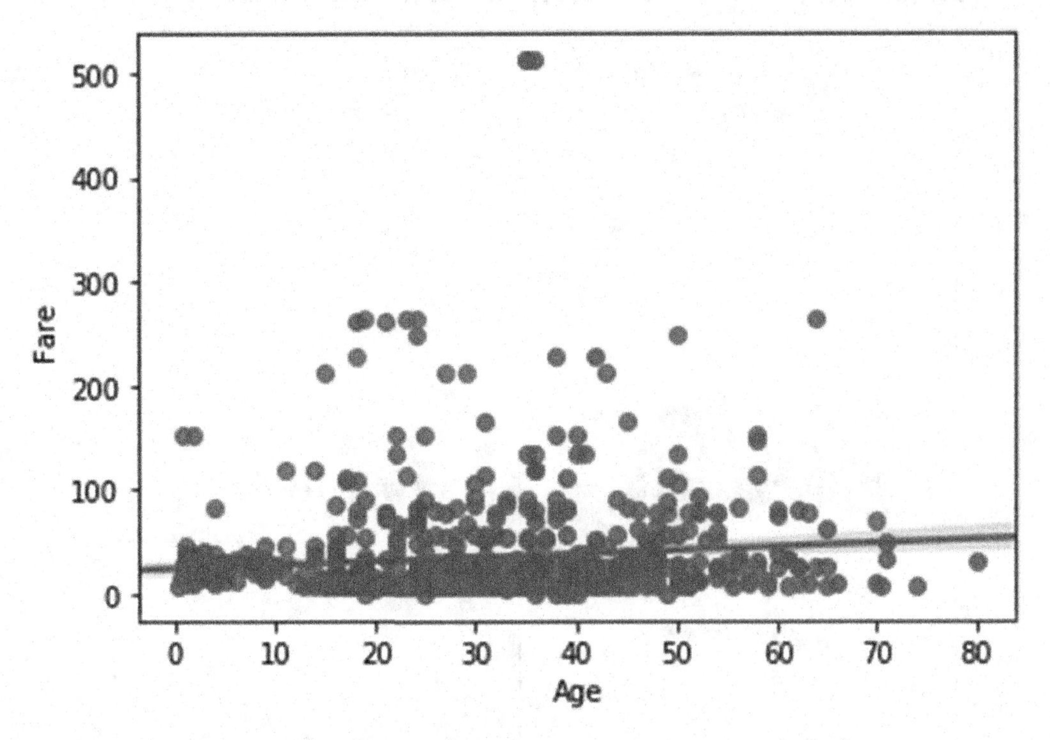

Figure 7-11. *An example of a regplot*

Further reading:

https://seaborn.pydata.org/generated/seaborn.regplot.html#seaborn.regplot

lmplot

This plot is a combination of a regplot and a facet grid, as shown in Figure 7-12. Using the *lmplot* function, we can see the relationship between two continuous variables across different parameter values.

In the following example, we plot two numeric variables ("Age" and "Fare") across a grid with different row and column variables.

CODE:

```
sns.lmplot(x='Age',y='Fare',row='Survived',data=titanic,col='Sex')
```

Output:

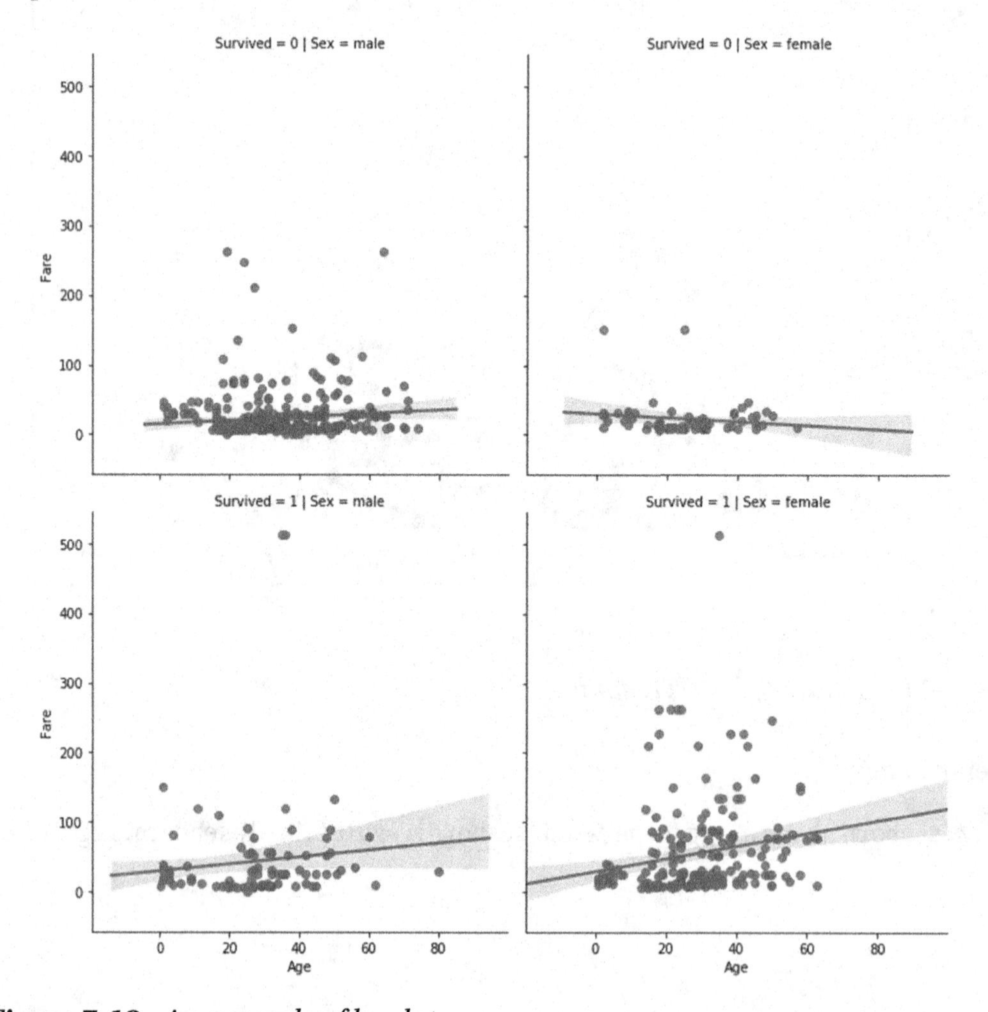

Figure 7-12. *An example of lmplot*

The following summarizes the differences between regplot and lmplot:

- The *regplot* function takes only two variables as arguments, whereas the lmplot function accepts multiple arguments.
- The *lmplot* function works at the level of the figure object, while the regplot function works at the level of an axes object.

Further reading on the lmplot: `https://seaborn.pydata.org/generated/seaborn.lmplot.html`

See more on the differences between regplot and lmplot: `https://seaborn.pydata.org/tutorial/regression.html#functions-to-draw-linear-regression-models`

Strip plot

A strip plot is similar to a scatter plot. The difference lies in the type of variables used in a strip plot. While a scatter plot has both variables as continuous, a strip plot plots one categorical variable against one continuous variable, as shown in Figure 7-13. The Seaborn function *striplot* generates a strip plot.

Consider the following example, where the "Age" variable is continuous, while the "Survived" variable is categorical.

CODE:

```
sns.stripplot(x='Survived',y='Age',data=titanic)
```

Output:

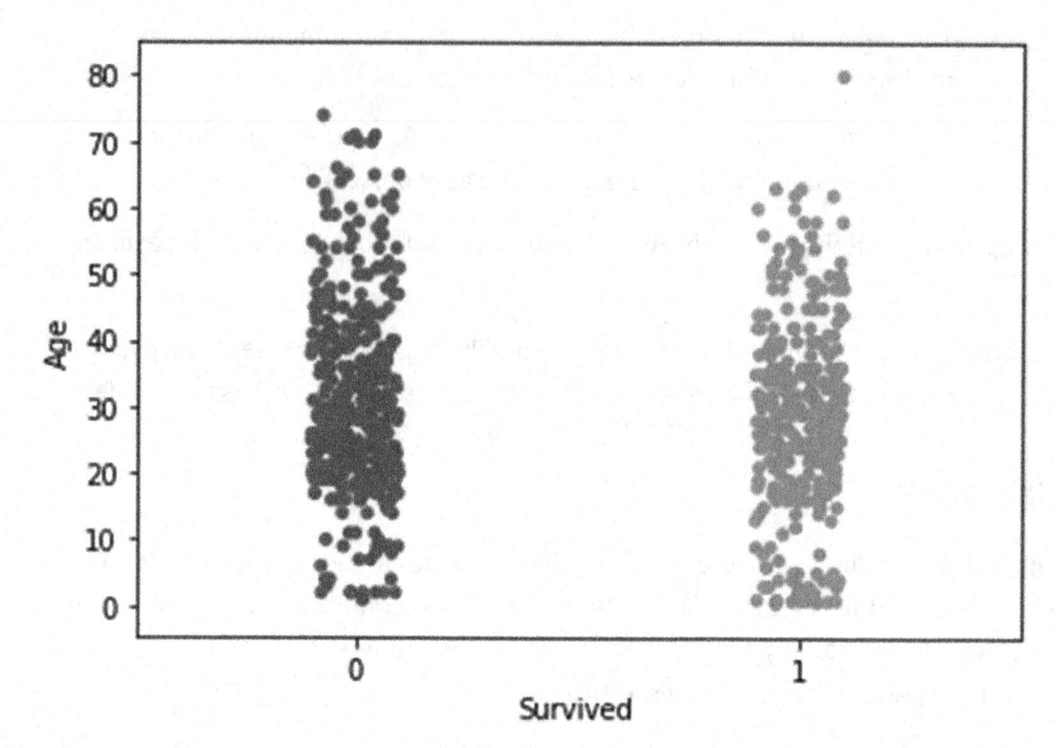

Figure 7-13. *An example of a strip plot*

Further reading: `https://seaborn.pydata.org/generated/seaborn.stripplot.html`

Swarm plot

A swarm plot is similar to a strip plot, the difference being that the points in a swarm plot are not overlapping like those in a strip plot. With the points more spread out, we get a better idea of the distribution of the continuous variable, as shown in Figure 7-14. The Seaborn function *swarmplot* generates a swarm plot.

CODE:

```
sns.swarmplot(x='Survived',y='Age',data=titanic)
```

Output:

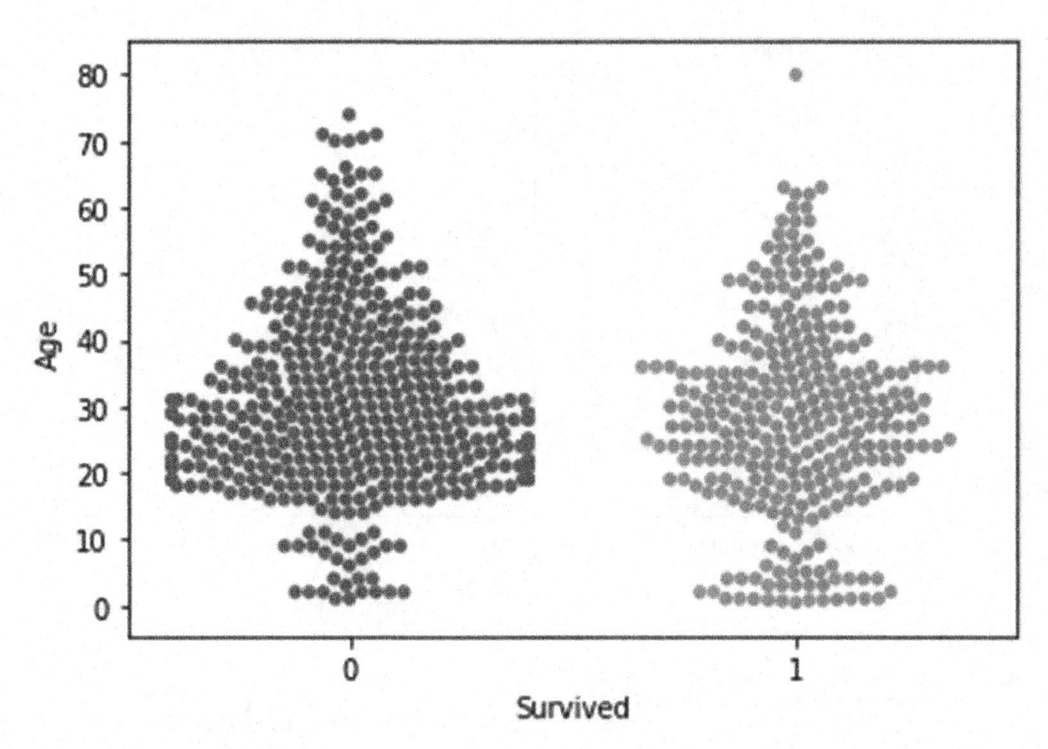

Figure 7-14. *An example of a swarm plot*

Further reading: `https://seaborn.pydata.org/generated/seaborn.swarmplot.html#seaborn.swarmplot`

Catplot

A catplot is a combination of a strip plot and a facet grid. We can plot one continuous variable against various categorical variables by specifying the *row, col,* or *hue* parameters, as shown in Figure 7-15. Note that while the strip plot is the default plot generated by the *catplot* function, it can generate other plots too. The type of plot can be changed using the *kind* parameter.

CODE:

```
sns.catplot(x='Survived',y='Age',col='Survived',row='Sex',data=titanic)
```

Output:

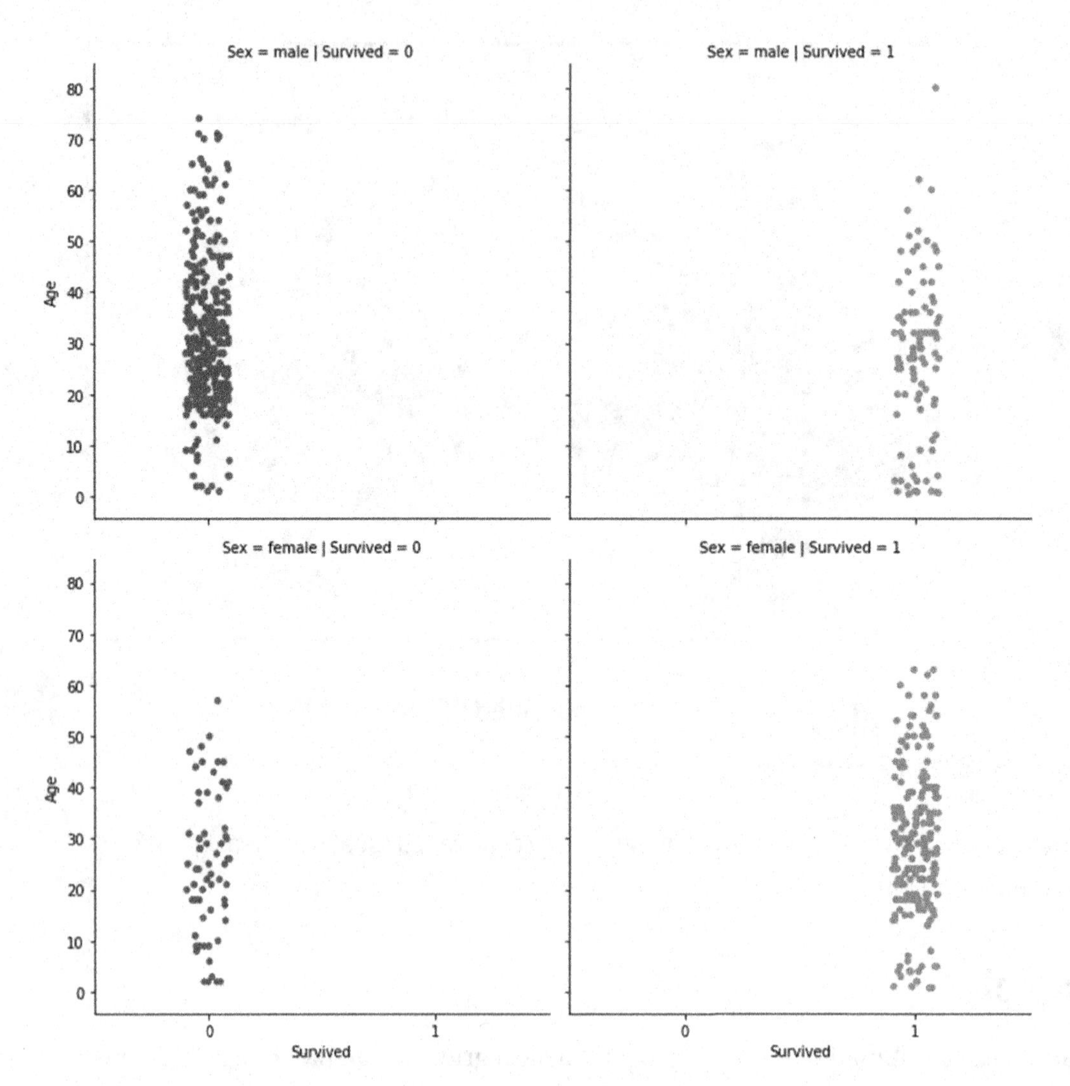

Figure 7-15. *An example of a catplot in Seaborn*

Further reading: `https://seaborn.pydata.org/generated/seaborn.catplot.html`

Pair plot

A pair plot is one that shows bivariate relationships between all possible pairs of variables in the dataset, as shown in Figure 7-16. The Seaborn function *pairplot* creates a pair plot. Notice that you do not have to supply any column names as arguments since all the variables in the dataset are considered automatically for plotting. The only parameter that you need to pass is the name of the DataFrame. In some of the plots displayed as part of the pair plot output, any given variable is also plotted against itself. The plots along the diagonal of a pair plot show these plots.

CODE:

```
sns.pairplot(data=titanic)
```

Output:

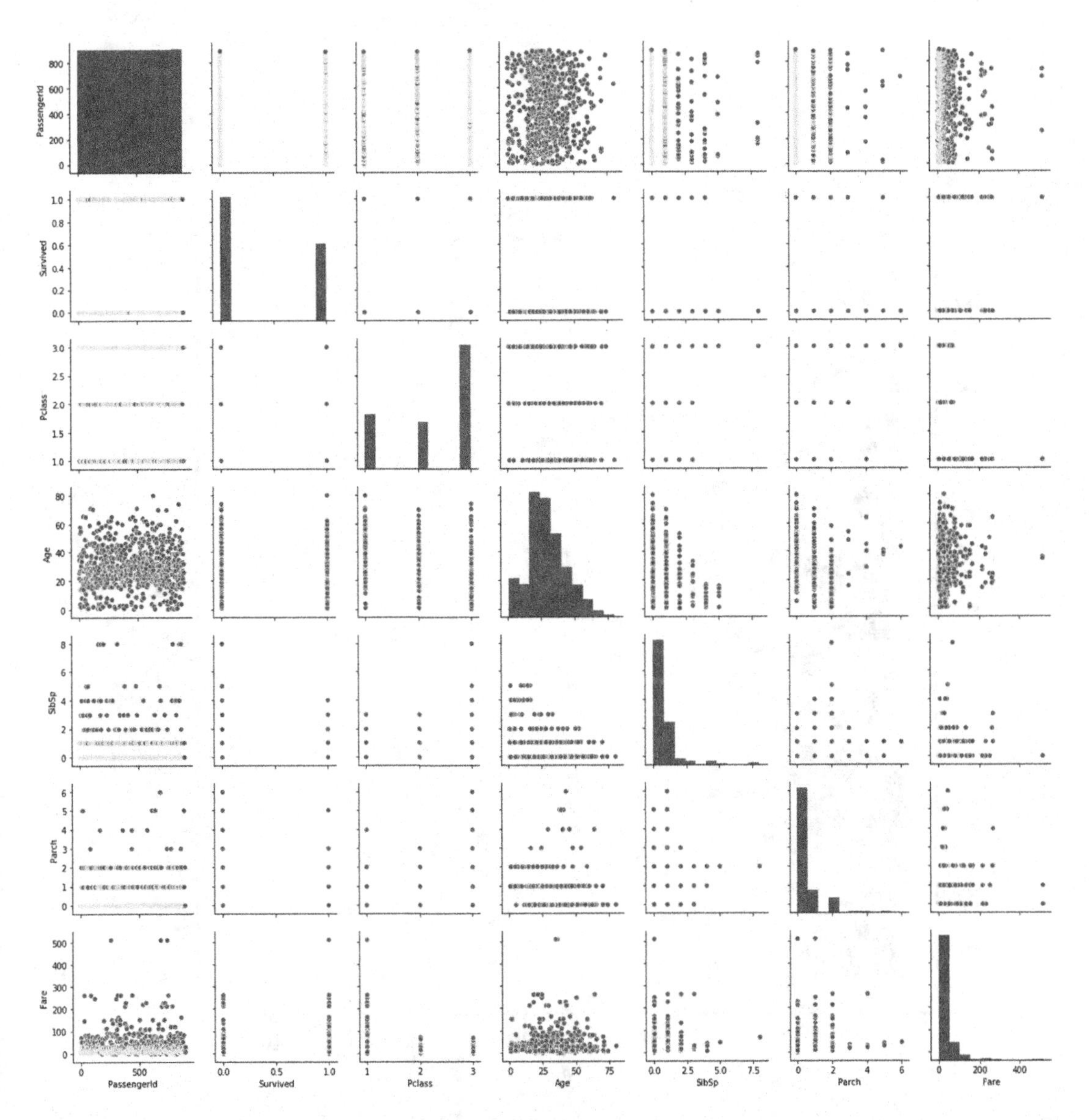

Figure 7-16. *An example of a pair plot in Seaborn*

Joint plot

The joint plot displays the relationship between two variables as well as the individual distribution of the variables, as shown in Figure 7-17. The *jointplot* function takes the names of the two variables to be plotted as arguments.

CODE:

```
sns.jointplot(x='Fare',y='Age',data=titanic)
```

Output:

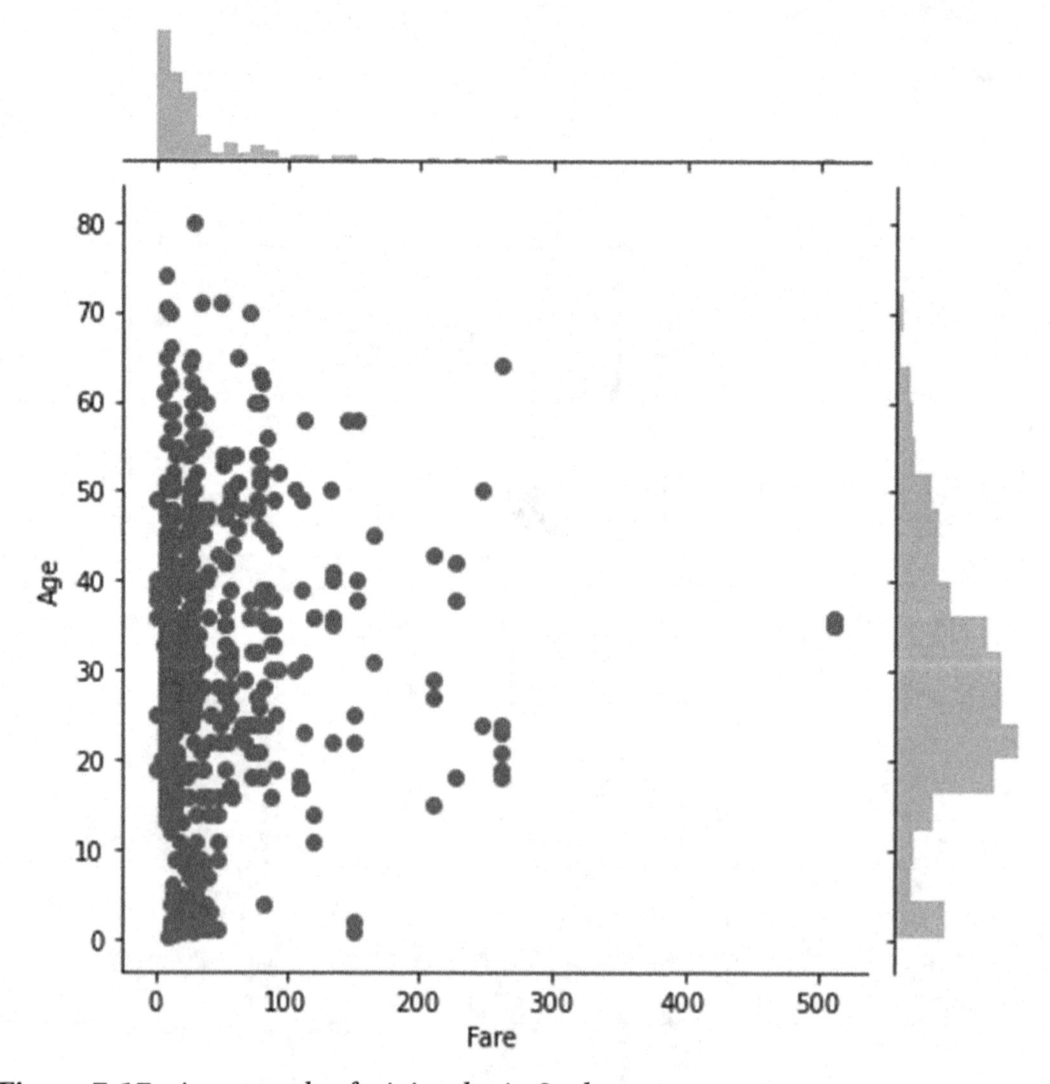

Figure 7-17. *An example of a jointplot in Seaborn*

Further reading on the jointplot: `https://seaborn.pydata.org/generated/seaborn.jointplot.html#seaborn.jointplot`

Further reading on the Seaborn library:

Examples of various graphs that can be created in Seaborn: `https://seaborn.pydata.org/examples/index.html`

Getting started and solved examples in Seaborn: `https://seaborn.pydata.org/introduction.html`

Summary

1. Three Python-based libraries can be used for visualization in Python – Matplotlib (which is based on Matlab), Pandas, and Seaborn.

2. Before we draw graphs, we need to figure out the type of variable that needs to be plotted and the number of variables that need to be plotted. Use bar charts and pie charts for categorical variables, and histograms and scatter plots for continuous variables.

3. Matplotlib has two interfaces that are used for plotting – the stateful interface and the object-oriented interface. The stateful interface uses the *pyplot* class and keeps track of the current state of the object of this class. The object-oriented interface uses a hierarchy of objects to represent various elements of the plot and uses these objects for plotting.

4. The *plot* function, which is used in Pandas for plotting, uses Matplotlib at the back end. This function makes it easy to draw any kind of graph just by changing the arguments passed to it, thus utilizing the principle of polymorphism (one name, many forms).

5. Seaborn is another library that uses Matplotlib at the back end. By changing its default parameters, it minimizes the need to perform aggregations, label, and color code the elements of your graph. It also provides the ability to visualize more than two variables.

In the next chapter, we examine some real-world case studies where we will put into practice what we have learned about visualization and data wrangling.

Review Exercises

Question 1

Which plot would you use when you have the following variables?

- One categorical variable
- One continuous variable
- Two continuous variables
- Three or more continuous variables
- One continuous and one categorical variable
- Two continuous and two or more categorical variables

Question 2

Match the functions on the left with the correct description on the right

1. Facet grid	a. Plot showing relationships between all possible pairs of variables and distributions of individual variables
2. Catplot	b. Plot of a continuous variable across a grid of categorical parameters
3. Swarm plot	c. Plot of one continuous and one categorical variable, with points not overlapping
4. Pair plot	d. Plot that combines a box plot with a kernel density estimate
5. Violin plot	e. Combination of a facet grid and strip plot

Question 3

Which among the following is true about the visualization performed using the Pandas library?

1. Aggregation is performed automatically while plotting graphs
2. Pandas requires data in the long format
3. The *plot* method is used to plot graphs
4. The *type* parameter is used to specify the type of plot
5. The *kind* parameter to specify the type of plot

Question 4

The magic command needed for displaying Matplotlib and Pandas graphs inline is

1. %matplotlib
2. %inline
3. %matplotlib inline
4. None of the above

Question 5

The axes object refers to the

1. x axis
2. y axis
3. Both the x and y axis
4. The subplot containing the graph

Question 6

For a given DataFrame, df, how do we specify the following parameters used in the heatmap function?

1. Correlation matrix
2. Color map for coloring the squares in a heatmap
3. The numeric value of the degree of correlation between each of the parameters

Question 7

The Sklearn library has a built-in dataset, *Iris*. It contains samples from various species of the iris plant. Each sample contains four attributes: sepal length, sepal width, petal length, petal width, and species (*Iris setosa, Iris versicolor,* and *Iris virginica*), and there are 50 samples of each species.

- Read the data from this dataset into a DataFrame.
- Create a 10*5 figure with two subplots.
- In the first subplot, plot the petal length vs. petal width.

- In the second subplot, plot the sepal length vs. sepal width.
- For each of the plots, label the x and y axes and set a title.

Question 8

Load the data from tips dataset (built into Seaborn) using the load_dataset function in Seaborn. This dataset contains the total bill amounts and the tip values for different customers visiting a restaurant. The customers are categorized according to their gender, smoking preference, and the day and time of their visit.

- Create a plot that shows the distribution (strip plot) of the total bill for smokers and nonsmokers, across a grid containing different values for the time and sex columns.
- Create a plot to show the relationship between the "tip" and "total_bill" columns for: males and smokers, males and nonsmokers, females and smokers, and females and nonsmokers.

Answers

Question 1

- One categorical variable: count plot
- One continuous variable: histogram, kernel density estimate
- Two continuous variables: scatter plot, line plot
- Three or more continuous variables: heat map
- One continuous and one categorical variable: strip plot, swarm plot, bar plot
- Two continuous and two or more categorical variables: catplot, facet grid, lmplot

Question 2

1-b; 2-e; 3-c; 4-a; 5-d

Question 3

Options 3 and 5

Pandas uses the *plot* method with the kind parameter to create various graphs. Pandas requires data to be in the wide or aggregated form. Aggregation is not done by default, unlike Seaborn. The *value_counts* method is required for aggregation before the *plot* method is applied.

Question 4

Option 3

We use the magic command (*%matplotlib inline*) for displaying graphs inline in Matplotlib and Pandas.

Question 5

Option 4

The term "axes" is a misnomer and does not refer to the x or y axis. It refers to the subplot or plotting area, which is a part of the figure object. A figure can contain multiple subplots.

Question 6

- Correlation matrix: df.corr()(generated a DataFrame representing the correlation matrix)
- Color map for coloring the squares in a heatmap: *cmap* parameter
- Denoting the numeric value of the degree of correlation: annot parameter (*annot=True*)

Question 7

```
import matplotlib.pyplot as plt
from sklearn.datasets import load_iris
import pandas as pd
#importing the iris dataset
data=load_iris()
iris=pd.DataFrame(data=data.data,columns=data.feature_names)
iris['species']=data.target
iris['species']=iris['species'].map({0:'setosa',1:'versicolor',2:'virginica'})
iris['species'].value_counts().plot(kind='bar')
fig=plt.figure(figsize=(10,5))
```

```
ax1=fig.add_subplot(121) #defining the first subplot
#plotting petal length vs petal width in the first subplot
iris.plot(kind='scatter',x='petal length (cm)',y='petal width (cm)',ax=ax1)
#adding the title for the first subplot
ax1.set_title("Petal length vs width")
#adding the label for the X axis
ax1.set_xlabel("Petal length")
#adding the label for the Y axis
ax1.set_ylabel("Petal width")
ax2=fig.add_subplot(122) #defining the second subplot
#plotting sepal length vs sepal width in the second subplot
iris.plot(kind='scatter',x='sepal length (cm)',y='sepal width (cm)',ax=ax2)
ax2.set_xlabel("Sepal length")
ax2.set_ylabel("Sepal width")
ax2.set_title("Sepal length vs width")
#Increasing the distance width between the subplots
fig.subplots_adjust(wspace=1)
```

Question 8

```
import seaborn as sns
#loading the data into a DataFrame using the load_dataset function
tips = sns.load_dataset("tips")
#creating a catplot for showing the distribution of the total bill for
different combinations of parameters
sns.catplot(x='smoker',y='total_bill',row='time',col='sex',data=tips)
#defining the FacetGrid object and setting the row and column values
g=sns.FacetGrid(tips,row='sex',col='smoker')
#specifying the plot and the columns we want to display
g.map(plt.scatter,'tip','total_bill')
```

CHAPTER 8

Data Analysis Case Studies

In the last chapter, we looked at the various Python-based visualization libraries and how the functions from these libraries can be used to plot different graphs. Now, we aim to understand the practical applications of the concepts we have discussed so far with the help of case studies. We examine the following three datasets:

- Analysis of unstructured data: Using data from a web page providing information about the top 50 highest-grossing movies in France during the year 2018
- Air quality analysis: Data from an air quality monitoring station at New Delhi (India), providing the daily levels for four pollutants – sulfur dioxide (SO_2), oxides of nitrogen as nitrogen dioxide (NO_2), ozone, and fine particulate matter ($PM_{2.5}$)
- COVID-19 trend analysis: Dataset capturing the number of cases and deaths for various countries across the world daily for the first six months in the year 2020

Technical requirements

External files

For the first case study, you need to refer to the following Wikipedia URL (data is taken directly from the web page):

```
https://en.wikipedia.org/wiki/List_of_2018_box_office_number-one_films_in_
France
```


G. Rajagopalan, *A Python Data Analyst's Toolkit*, https://doi.org/10.1007/978-1-4842-6399-0_8

For the second case study, download a CSV file from the following link:

`https://github.com/DataRepo2019/Data-files/blob/master/NSIT%20Dwarka.csv`

For the third case study, download an Excel file from the following link: `https://github.com/DataRepo2019/Data-files/blob/master/COVID-19-geographic-disbtribution-worldwide-2020-06-29.xlsx`

Libraries

In addition to the modules and libraries we used in the previous chapters (including Pandas, NumPy, Matplotlib, and Seaborn), we use the *requests* module in this chapter to make HTTP requests to websites.

To use the functions contained in this module, import this module in your Jupyter notebook using the following line:

```
import requests
```

If the *requests* modules is not installed, you can install it using the following command on the Anaconda Prompt.

```
pip install requests
```

Methodology

We will be using the following methodology for each of the case studies:

1. Open a new Jupyter notebook, and perform the following steps:
    - Import the libraries and data necessary for your analysis
    - Read the dataset and examine the first five rows (using the *head* method)
    - Get information about the data type of each column and the number of non-null values in each column (using the *info* method) and the dimensions of the dataset (using the *shape* attribute)
    - Get summary statistics for each column (using the *describe* method) and obtain the values of the count, min, max, standard deviation, and percentiles

2. Data wrangling
    - Check if the data types of the columns are correctly identified (using the *info* or *dtype* method). If not, change the data types, using the *astype* method
    - Rename the columns if necessary, using the *rename* method
    - Drop any unnecessary or redundant columns or rows, using the *drop* method
    - Make the data tidy, if needed, by restructuring it using the *melt* or *stack* method
    - Remove any extraneous data (blank values, special characters, etc.) that does not add any value, using the *replace* method
    - Check for the presence of null values, using the *isna* method, and drop or fill the null values using the *dropna* or *fillna* method
    - Add a column if it adds value to your analysis
    - Aggregate the data if the data is in a disaggregated format, using the groupby method
3. Visualize the data using univariate, bivariate, and multivariate plots
4. Summarize your insights, including observations and recommendations, based on your analysis

Case study 8-1: Highest grossing movies in France – analyzing unstructured data

In this case study, the data is read from an HTML page instead of a conventional CSV file.

The URL that we are going to use is the following: `https://en.wikipedia.org/wiki/List_of_2018_box_office_number-one_films_in_France`

This web page has a table that displays data about the top 50 films in France by revenue, in the year 2018. We import this data in Pandas using methods from the Requests library. Requests is a Python library used for making HTTP requests. It helps with extracting HTML from web pages and interfacing with APIs.

Questions that we want to answer through our analysis:

1. Identify the top five films by revenue
2. What is the percentage share (revenue) of each of the top ten movies?
3. How did the monthly average revenue change during the year?

Step 1: Importing the data and examining the characteristics of the dataset

First, import the libraries and use the necessary functions to retrieve the data.

CODE:

```
#importing the libraries
import requests
import pandas as pd
import seaborn as sns
import matplotlib.pyplot as plt
#Importing the data from the webpage into a DataFrame
url='https://en.wikipedia.org/wiki/List_of_2018_box_office_number-one_
films_in_France'
req=requests.get(url)
data=pd.read_html(req.text)
df=data[0]
```

We import all the libraries and store the URL in a variable. Then we make an HTTP request to this URL using the get method to retrieve information from this web page. The text attribute of the requests object contains the HTML data, which is passed to the *pd.read_html* function. This function returns a list of DataFrame objects containing the various tables on the web page. Since there is only one table on the web page, the DataFrame (df) contains only one table.

Examining the first few records:

CODE:

```
df.head()
```

Output:

	#	Date	Film	Gross	Notes
0	1	January 7, 2018	Star Wars: The Last Jedi	US$6,557,062	[1]
1	2	January 14, 2018	Jumanji: Welcome to the Jungle	US$2,127,871	[2]
2	3	January 21, 2018	Brillantissime	US$2,006,033	[3]
3	4	January 28, 2018	The Post	US$2,771,269	[4]
4	5	February 4, 2018	Les Tuche 3	US$16,604,101	[5]

Obtaining the data types and missing values:

CODE:

```
df.info()
```

Output:

```
<class 'pandas.core.frame.DataFrame'>
RangeIndex: 50 entries, 0 to 49
Data columns (total 5 columns):
#        50 non-null int64
Date     50 non-null object
Film     50 non-null object
Gross    50 non-null object
Notes    50 non-null object
dtypes: int64(1), object(4)
memory usage: 2.0+ KB
```

As we can see, the data types of the columns are not in the format we need. The "Gross" column represents the gross revenue, which is a numeric column. This column, however, has been assigned an object data type because it contains numeric as well as non-numeric data (characters like ",", "$" symbol, and letters like "U" and "S"). In the next step, we deal with problems such as these.

Step 2: Data wrangling

In this step, we will:

1. Remove unnecessary characters
2. Change data types

3. Remove columns that are not needed

4. Create a new column from an existing column

Let us remove the unwanted strings from the "Gross" column, retaining only the numeric values:

CODE:

```
#removing unnecessary characters from the Gross column
df['Gross']=df['Gross'].str.replace(r"US\$","").str.replace(r",","")
```

In the preceding statement, we use a series of chained replace methods and the principle of regular expressions to replace the non-numeric characters. The first replace method removes "US$" and the second replace method removes the commas. Replacing a character with an empty string ("") is equivalent to removing the character.

Now, let us use the *astype* method to typecast or change the data type of this column to *int64* so that this column can be used for computations and visualizations:

CODE:

```
#changing the data type of the Gross column to make the column numeric
df['Gross']=df['Gross'].astype('int64')
```

To check whether these changes have been reflected, we examine the first few records of this column and verify the data type:

CODE:

```
df['Gross'].head(5)
```

Output:

```
0      6557062
1      2127871
2      2006033
3      2771269
4     16604101
Name: Gross, dtype: int64
```

As we can see from the output, the data type of this column has been changed, and the values do not contain strings any longer.

We also need to extract the month part of the date, which we will do by first changing the data type of the "Date" column and then applying the *DatetimeIndex* method to it, as shown in the following.

CODE:

```
#changing the data type of the Date column to extract its components
df['Date']=df['Date'].astype('datetime64')
#creating a new column for the month
df['Month']=pd.DatetimeIndex(df['Date']).month
```

Finally, we remove two unnecessary columns from the DataFrame, using the following statement.

CODE:

```
#dropping the unnecessary columns
df.drop(['#','Notes'],axis=1,inplace=True)
```

Step 3: Visualization

To visualize our data, first we create another DataFrame (df1), which contains a subset of the columns the original DataFrame (df) contains. This DataFrame, df1, contains just two columns – "Film" (the name of the movie) and "Gross" (the gross revenue). Then, we sort the values of the revenue in the descending order. This is shown in the following step.

CODE:

```
df1=df[['Film','Gross']].sort_values(ascending=False,by='Gross')
```

There is an unwanted column ("index") that gets added to this DataFrame that we will remove in the next step.

CODE:

```
df1.drop(['index'],axis=1,inplace=True)
```

Top Five Films:

The first plot we create is a bar graph showing the top five films in terms of revenue: (Figure 8-1).

```
#Plotting the top 5 films by revenue
#setting the figure size
plt.figure(figsize=(10,5))
#creating a bar plot
```

```
ax=sns.barplot(x='Film',y='Gross',data=df1.head(5))
#rotating the x axis labels
ax.set_xticklabels(labels=df1.head()['Film'],rotation=75)
#setting the title
ax.set_title("Top 5 Films per revenue")
#setting the Y-axis labels
ax.set_ylabel("Gross revenue")
#Labelling the bars in the bar graph
for p in ax.patches:

ax.annotate(p.get_height(),(p.get_x()+p.get_width()/2,p.get_height()),
ha='center',va='bottom')
```

Output:

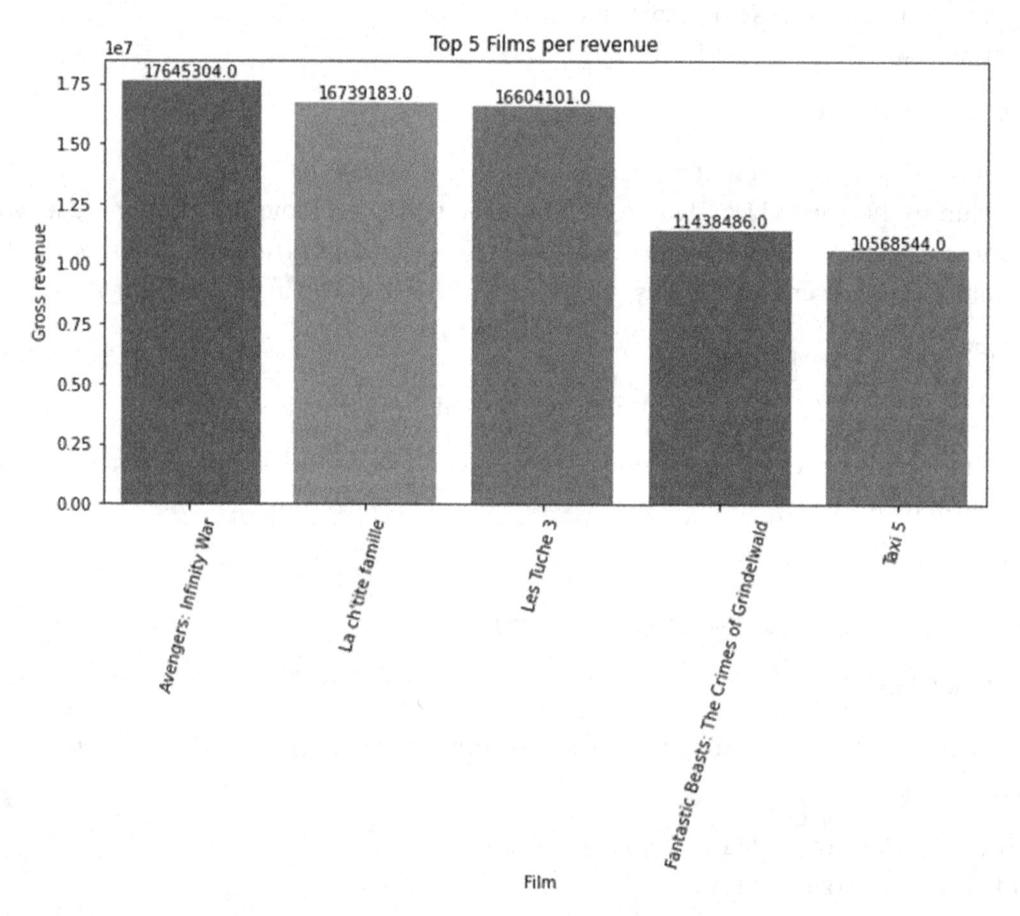

***Figure 8-1.** Pie chart showing the individual percentage share of each of the top five films*

To depict the share of the top ten films (by revenue), we create a pie chart (Figure 8-2).

CODE:

```
#Pie chart showing the share of each of the top 10 films by percentage in
the revenue
df1['Gross'].head(10).plot(kind='pie',autopct='%.2f%%',labels=df1['Film'],
figsize=(10,5))
```

Output:

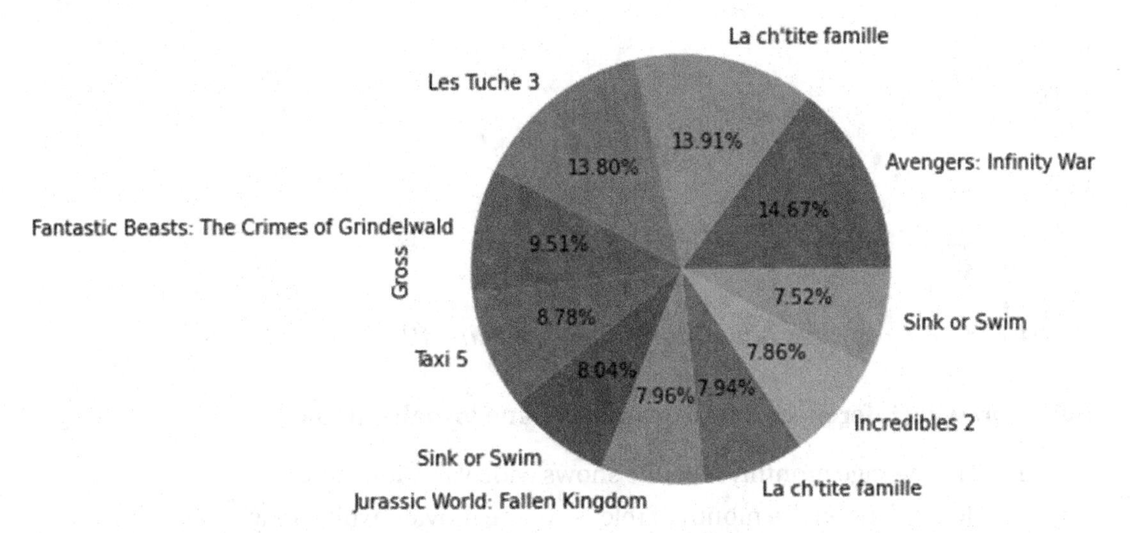

Figure 8-2. *Pie chart depicting the share of top ten movies by revenue*

We first create another DataFrame that aggregates the data for a month by calculating an average for each month (Figure 8-3).

CODE:

```
#Aggregating the revenues by month
df2=df.groupby('Month')['Gross'].mean()
#creating a line plot
df2.plot(kind='line',figsize=(10,5))
```

Output:

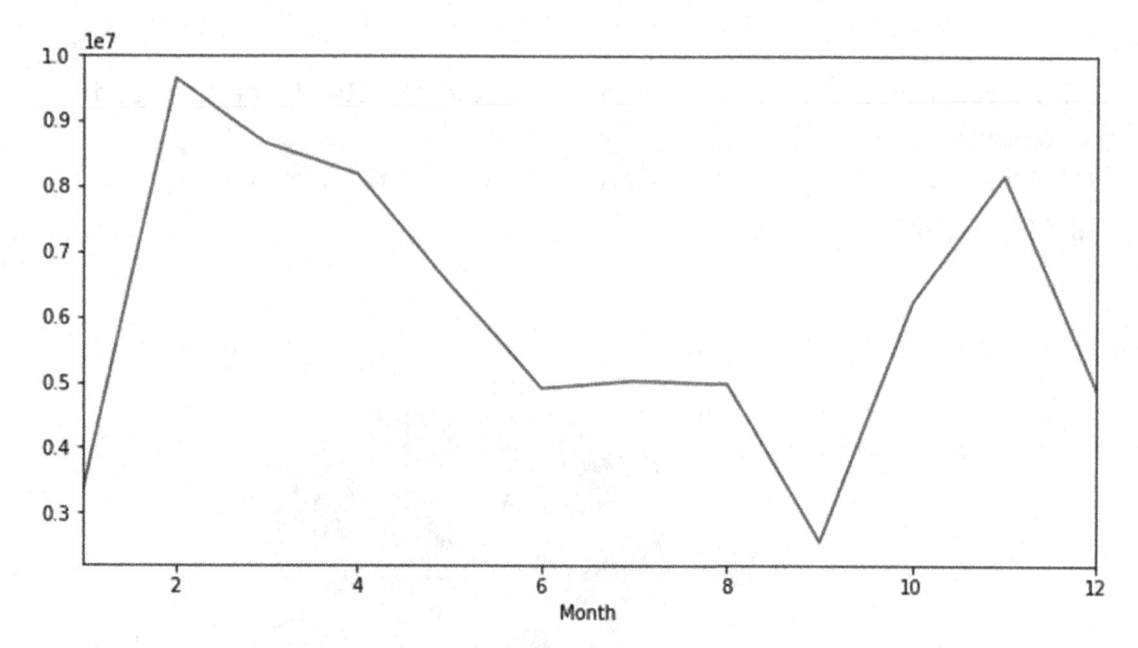

Figure 8-3. *Average box office monthly revenue in 2018 (France)*

Step 4: Drawing inferences based on analysis and visualizations

1. The average monthly revenue shows wide variation, possibly dependent on the month of release of the movies, which may necessitate further analysis across the years.

2. The top three highest-revenue-grossing movies in France in the year 2018 were *Avengers, La Ch'tite Famille,* and *Les Tuche 3.*

Case study 8-2: Use of data analysis for air quality management

To monitor the status of ambient air quality, The Central Pollution Control Board (CPCB), India, operates a vast network of monitoring stations spread across the country. Parameters regularly monitored include sulfur dioxide (SO_2), oxides of nitrogen as nitrogen dioxide (NO_2), ozone, and fine particulate matter ($PM_{2.5}$). Based on trends over the years, air quality in the national capital of Delhi has emerged as a matter of public concern. A stepwise analysis of daily air quality data follows to demonstrate how data analysis could assist in planning interventions as part of air quality management.

Note: The name of the dataset used for this case study is: "NSIT Dwarka.csv". Please refer to the technical description section for details on how to import this dataset.

Questions that we want to answer through our analysis:

1. Yearly averages: Out of the four pollutants - SO_2, NO_2, ozone, and $PM_{2.5}$ - which have yearly average levels that regularly surpass the prescribed annual standards?
2. Daily standards: For the pollutants of concern, on how many days in each year are the daily standards exceeded?
3. Temporal variation: Which are the months where the pollution levels exceed critical levels on most days?

Step 1: Importing the data and examining the characteristics of the dataset

CODE:

```
import pandas as pd
import numpy as np
import seaborn as sns
import matplotlib.pyplot as plt
#aqdata is the name of the DataFrame, short for Air Quality Data.
aqdata=pd.read_csv('NSIT Dwarka.csv')
aqdata.head()
```

Output:

	From Date	To Date	PM2.5	SO2	Ozone	NO2
0	01-01-2014 00:00	02-01-2014 00:00	None	22.7	8.63	5.59
1	02-01-2014 00:00	03-01-2014 00:00	None	8.72	8.43	3.68
2	03-01-2014 00:00	04-01-2014 00:00	None	13.83	9.77	3.83
3	04-01-2014 00:00	05-01-2014 00:00	None	27.64	6.83	9.64
4	05-01-2014 00:00	06-01-2014 00:00	None	37.17	7.34	11.06

Checking for the data types of the columns:

CODE:

```
aqdata.info()
```

Output:

```
<class 'pandas.core.frame.DataFrame'>
RangeIndex: 2192 entries, 0 to 2191
Data columns (total 6 columns):
From Date     2191 non-null object
To Date       2191 non-null object
PM2.5         2191 non-null object
SO2           2187 non-null object
Ozone         2187 non-null object
NO2           2190 non-null object
dtypes: object(6)
memory usage: 102.8+ KB
```

Observation: Even though the values for SO_2, NO_2, ozone, and $PM_{2.5}$ are numeric, Pandas reads the data type of these columns as "*object*". To work with these columns (i.e., plot graphs, observe trends, calculate aggregate values), we need to change the data types of these columns. Also, there seem to be some missing entries.

Step 2: Data wrangling

Based on the observations in the previous step, in this step, we will

1. Deal with missing values: We have the option of either dropping the null values or substituting the null values.
2. Change the data types for the columns.

Checking for missing values in the dataset:

CODE:

```
aqdata.isna().sum()
```

Output:

```
From Date     1
To Date       1
PM2.5         1
SO2           5
Ozone         5
NO2           2
dtype: int64
```

There do not seem to be many missing values, but herein lies the catch. When we examined the first few rows using the head statement, we saw that some missing values are represented as *None* in the original dataset. However, these are not being recognized as null values by Pandas. Let us replace the value, *None*, with the value *np.nan* so that Pandas acknowledges these values as null values:

CODE:

```
aqdata=aqdata.replace({'None':np.nan})
```

Now, if we count the number of null values, we see a vastly different picture, indicating a much higher presence of missing values in the dataset.

CODE:

```
aqdata.isna().sum()
```

Output:

```
From Date       1
To Date         1
PM2.5         562
SO2            84
Ozone         106
NO2           105
dtype: int64
```

Let us check the current data types of the columns:

CODE:

```
aqdata.info()
```

Output:

```
<class 'pandas.core.frame.DataFrame'>
RangeIndex: 2192 entries, 0 to 2191
Data columns (total 6 columns):
From Date    2191 non-null object
To Date      2191 non-null object
PM2.5        1630 non-null object
SO2          2108 non-null object
Ozone        2086 non-null object
NO2          2087 non-null object
dtypes: object(6)
memory usage: 102.8+ KB
```

We see that the columns containing numeric values are not recognized as numeric columns, and the columns containing dates are also not recognized correctly. Having columns with incorrect data types becomes an impediment for the next step, where we analyze trends and plot graphs; this step requires the data types of the columns to be in a format that is appropriate for Pandas to read.

In the following lines of code, we use the *pd.to_datetime* method to convert the data type of the "From Date" and "To Date" columns to the *datetime* type, which makes it easier to analyze individual components of the date like months and years.

CODE:

```
aqdata['From Date']=pd.to_datetime(aqdata['From Date'], format='%d-%m-%Y %H:%M')
aqdata['To Date']=pd.to_datetime(aqdata['To Date'], format='%d-%m-%Y %H:%M')
aqdata['SO2']=pd.to_numeric(aqdata['SO2'],errors='coerce')
aqdata['NO2']=pd.to_numeric(aqdata['NO2'],errors='coerce')
aqdata['Ozone']=pd.to_numeric(aqdata['Ozone'],errors='coerce')
aqdata['PM2.5']=pd.to_numeric(aqdata['PM2.5'],errors='coerce')
```

Use the *info* method to check whether the data types have been changed.

CODE:

```
aqdata.info()
```

Output:

```
<class 'pandas.core.frame.DataFrame'>
RangeIndex: 2192 entries, 0 to 2191
Data columns (total 6 columns):
From Date    2191 non-null datetime64[ns]
To Date      2191 non-null datetime64[ns]
PM2.5        1630 non-null float64
SO2          2108 non-null float64
Ozone        2086 non-null float64
NO2          2087 non-null float64
dtypes: datetime64[ns](2), float64(4)
memory usage: 102.8 KB
```

Since most of our analysis considers yearly data, we create a new column to extract the year, using the *pd.DatetimeIndex* function.

CODE:

```
aqdata['Year'] = pd.DatetimeIndex(aqdata['From Date']).year
```

Now, we create separate DataFrame objects for each year so that we can analyze the data yearly.

CODE:

```
#extracting the data for each year
aq2014=aqdata[aqdata['Year']==2014]
aq2015=aqdata[aqdata['Year']==2015]
aq2016=aqdata[aqdata['Year']==2016]
aq2017=aqdata[aqdata['Year']==2017]
aq2018=aqdata[aqdata['Year']==2018]
aq2019=aqdata[aqdata['Year']==2019]
```

Now, let us have a look at the number of null values in the data for each year:

CODE:

```
#checking the missing values in 2014
aq2014.isna().sum()
```

Output:

```
From Date      0
To Date        0
PM2.5        365
SO2            8
Ozone          8
NO2            8
Year           0
dtype: int64
```

```
#checking the missing values in 2015
aq2015.isna().sum()
```

Output:

```
From Date      0
To Date        0
PM2.5        117
SO2           12
Ozone         29
NO2           37
Year           0
dtype: int64
```

CODE:

```
#checking the missing values in 2016
aq2016.isna().sum()
```

Output:

```
From Date     0
To Date       0
PM2.5        43
SO2          43
Ozone        47
NO2          42
Year          0
dtype: int64
```

```
#checking the missing values in 2017
aq2017.isna().sum()
```

Output:

```
From Date     0
To Date       0
PM2.5        34
SO2          17
Ozone        17
NO2          12
Year          0
dtype: int64
```

CODE:

```
#checking the missing values in 2018
aq2018.isna().sum()
```

Output:

```
From Date    0
To Date      0
PM2.5        2
SO2          2
Ozone        2
NO2          2
Year         0
dtype: int64
```

CODE:

```
#checking the missing values in 2019
aq2019.isna().sum()
```

Output:

```
From Date    0
To Date      0
PM2.5        0
SO2          1
Ozone        2
NO2          3
Year         0
dtype: int64
```

From the analysis of null values for each year, we see that data for the years 2014 and 2015 have the majority of the missing values. Hence, we choose to disregard data from the years 2014 and 2015, and analyze data for 4 years from 2016 to 2019. As per norms set by the Central Pollution Control Board, India, we need at least 104 daily monitored values to arrive at annual averages.

2016, 2017, 2018, and 2019 are the four years for which air quality data would be analyzed. Before moving on to the next step, we drop the missing values for each year from 2016 to 2019 instead of substituting them since we have sufficient data (more than 104 readings) for each of these four years to calculate annual averages, as shown below.

CODE:

```
#dropping the null values for the four years chosen for analysis
aq2016.dropna(inplace=True) # inplace=True makes changes in the original
dataframe
aq2017.dropna(inplace=True)
aq2018.dropna(inplace=True)
aq2019.dropna(inplace=True)
```

Step 3: Data visualization

Part 1 of analysis: Plotting the yearly averages of the pollutants

Based on monitored 24-hourly average ambient air concentrations of $PM_{2.5}$, SO_2, NO_2, and ozone (O_3), yearly averages are plotted to identify parameters for which the prescribed national ambient air quality standards for annual average are exceeded.

First, we calculate the yearly averages for each pollutant ($PM_{2.5}$, SO_2, NO_2, and ozone), as follows:

CODE:

```
#Yearly averages for SO2 in each year
s16avg=round(aq2016['SO2'].mean(),2)
s17avg=round(aq2017['SO2'].mean(),2)
s18avg=round(aq2018['SO2'].mean(),2)
s19avg=round(aq2019['SO2'].mean(),2)
#Yearly averages for PM2.5 in each year
p16avg=round(aq2016['PM2.5'].mean(),2)
p17avg=round(aq2017['PM2.5'].mean(),2)
```

```
p18avg=round(aq2018['PM2.5'].mean(),2)
p19avg=round(aq2019['PM2.5'].mean(),2)
#Yearly averages for NO2 in each year
n16avg=round(aq2016['NO2'].mean(),2)
n17avg=round(aq2017['NO2'].mean(),2)
n18avg=round(aq2018['NO2'].mean(),2)
n19avg=round(aq2019['NO2'].mean(),2)
```

Explanation: The notation for naming variables representing the averages of pollutants is as follows: the first letter of the pollutant, the year, and the abbreviation "avg" for average. For instance, s15avg denotes the average level of SO_2 in the year 2015. We use the *mean* method to calculate the average and the round function to round the average value to two decimal points. We do not consider ozone since yearly standards do not apply to ozone.

Next, we create a DataFrame for each pollutant with two columns each. One of the columns represents the year, and the other column shows the yearly average level for that year.

CODE:

```
#Creating data frames with yearly averages for each pollutant
dfs=pd.DataFrame({'Yearly average':[s16avg,s17avg,s18avg,s19avg]},ind
ex=['2016','2017','2018','2019']) #dfs is for SO2
dfp=pd.DataFrame({'Yearly average':[p16avg,p17avg,p18avg,p19avg]},ind
ex=['2016','2017','2018','2019']) #dfp is for PM2.5
dfn=pd.DataFrame({'Yearly average':[n16avg,n17avg,n18avg,n19avg]},ind
ex=['2016','2017','2018','2019']) #dfn is for NO2
```

Now, we are ready to plot the graphs for the yearly averages of each pollutant (Figure 8-4).

CODE:

```
#Creating a figure with 3 subplots - 1 for each pollutant
fig,(ax1,ax2,ax3)=plt.subplots(1,3)
#Creating a DataFrame the yearly averages for NO2
dfn.plot(kind='bar',figsize=(20,5),ax=ax1)
#Setting the title for the first axes object
ax1.set_title("NO2", fontsize=18)
```

```
#Setting the X-axis label for the NO2 graph
ax1.set_xlabel("Years", fontsize=18)
ax1.legend().set_visible(False)
#Setting the Y-axis label
ax1.set_ylabel("Yearly average", fontsize=18)
#Creating a dashed line to indicate the annual standard
ax1.hlines(40, -.9,15, linestyles="dashed")
#Labelling this dashed line
ax1.annotate('Annual avg. standard for NO2',(-0.5,38))
#labelling the bars
for p in ax1.patches:
    ax1.annotate(p.get_height(),(p.get_x()+p.get_width()/2,p.get_height()),
    color="black", ha="left", va ='bottom',fontsize=12)

#Plotting the yearly averages similarly for PM2.5
dfp.plot(kind='bar',figsize=(20,5),ax=ax2)
ax2.set_title("PM2.5", fontsize=18)
ax2.hlines(40, -.9,15, linestyles="dashed")
ax2.annotate('Annual avg. standard for PM2.5',(-0.5,48))
ax2.legend().set_visible(False)
for p in ax2.patches:
    ax2.annotate(p.get_height(),(p.get_x()+p.get_width()/2,p.get_height()),
    color="black", ha="center", va ='bottom',fontsize=12)

#Plotting the yearly averages similarly for SO2
dfs.plot(kind='bar',figsize=(20,5),ax=ax3)
ax3.hlines(50, -.9,15, linestyles="dashed")
ax3.annotate('Annual avg. standard for SO2',(-0.5,48))
ax3.set_title("SO2", fontsize=18)
ax3.legend().set_visible(False)
for p in ax3.patches:
    ax3.annotate(p.get_height(),(p.get_x()+p.get_width()/2,p.get_height()),
    color="black", ha="center", va ='bottom',fontsize=12)
```

Output:

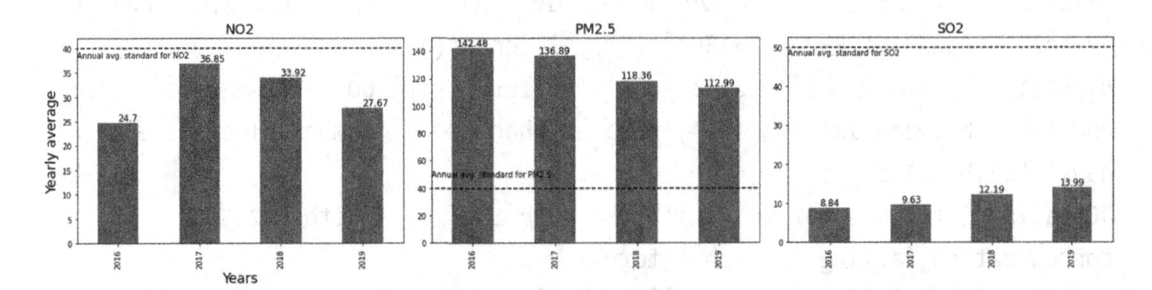

Figure 8-4. *The yearly average level for the pollutants (NO_2, $PM_{2.5}$, and SO_2), vis-à-vis their annual average standard*

Observation: It is evident that standards for annual average are exceeded only for $PM_{2.5}$. For NO_2, the observed values are relatively close to the prescribed standard. For SO_2, the observed values are much less than the annual standard. Therefore, for further analysis, only these two pollutants (NO_2 and $PM_{2.5}$) are considered.

Part 2 of air quality analysis: Plotting the number of days in each year where 24-hourly standards are exceeded for $PM_{2.5}$ and NO_2

While step 1 of the analysis indicates pollutants of concern for air quality management and planning of interventions, in step 2, for each year, we show how various levels of exceedance above standards for 24-hourly values are distributed. In the case of $PM_{2.5}$, we plot the number of days in each year for which observed values fall in the following ranges.

a. 0 to 60 μg/m^3

b. 61 to 120 μg/m^3

c. 121 to 180 μg/m^3

d. > 180 μg/m^3

To plot this data, we need to create DataFrame objects for each year from 2016 to 2019 capturing the number of days with $PM_{2.5}$ levels falling in each of these intervals, as shown in the following:

CODE:

```
#Creating intervals for 2016 with the number of days with PM2.5
concentration falling in that interval
a2=aq2016[(aq2016['PM2.5']<=60)]['PM2.5'].count()
```

```
b2=aq2016[((aq2016['PM2.5']>60) & (aq2016['PM2.5']<=120))]['PM2.5'].count()
c2=aq2016[((aq2016['PM2.5']>120) & (aq2016['PM2.5']<=180))]['PM2.5'].count()
d2=aq2016[(aq2016['PM2.5']>180)]['PM2.5'].count()
dfpb2016=pd.DataFrame({'year':'2016','pm levels':['<60','between 61
and 120','between 121 and 180','greater than 180'],'number of critical
days':[a2,b2,c2,d2]})
#Creating intervals for 2017 with the number of days with PM2.5
concentration falling in each interval
a3=aq2017[(aq2017['PM2.5']<=60)]['PM2.5'].count()
b3=aq2017[((aq2017['PM2.5']>60) & (aq2017['PM2.5']<=120))]['PM2.5'].count()
c3=aq2017[((aq2017['PM2.5']>120) & (aq2017['PM2.5']<=180))]['PM2.5'].count()
d3=aq2017[(aq2017['PM2.5']>180)]['PM2.5'].count()
dfpb2017=pd.DataFrame({'year':'2017','pm levels':['<60','between 61
and 120','between 121 and 180','greater than 180'],'number of critical
days':[a3,b3,c3,d3]})
#Creating intervals for 2018 with the number of days with PM2.5
concentration falling in each interval
a4=aq2018[(aq2018['PM2.5']<=60)]['PM2.5'].count()
b4=aq2018[((aq2018['PM2.5']>60) & (aq2018['PM2.5']<=120))]['PM2.5'].count()
c4=aq2018[((aq2018['PM2.5']>120) & (aq2018['PM2.5']<=180))]['PM2.5'].count()
d4=aq2018[(aq2018['PM2.5']>180)]['PM2.5'].count()
dfpb2018=pd.DataFrame({'year':'2018','pm levels':['<60','between 61
and 120','between 121 and 180','greater than 180'],'number of critical
days':[a4,b4,c4,d4]})
#Creating intervals for 2019 with the number of days with PM2.5
concentration falling in each interval
a5=aq2019[(aq2019['PM2.5']<=60)]['PM2.5'].count()
b5=aq2019[((aq2019['PM2.5']>60) & (aq2019['PM2.5']<=120))]['PM2.5'].count()
c5=aq2019[((aq2019['PM2.5']>120) & (aq2019['PM2.5']<=180))]['PM2.5'].
count()
d5=aq2019[(aq2019['PM2.5']>180)]['PM2.5'].count()
dfpb2019=pd.DataFrame({'year':'2019','pm levels':['<60','between 61
and 120','between 121 and 180','greater than 180'],'number of critical
days':[a5,b5,c5,d5]})
```

Now, we plot a stacked bar chart for each year with these intervals. To do so, we need to create pivot tables as follows:

CODE:

```
dfpivot2019=dfpb2019.pivot(index='year',columns='pm levels',values='number
of critical days')
dfpivot2018=dfpb2018.pivot(index='year',columns='pm levels',values='number
of critical days')
dfpivot2017=dfpb2017.pivot(index='year',columns='pm levels',values='number
of critical days')
dfpivot2016=dfpb2016.pivot(index='year',columns='pm levels',values='number
of critical days')
```

Using these pivot tables, we create stacked bar charts (Figure 8-5) as follows:

CODE:

```
#Creating a figure with 4 sub-plots, one for each year from 2016-19
fig,(ax1,ax2,ax3,ax4)=plt.subplots(1,4)
fig.suptitle("Number of days per year in each interval")
cmp=plt.cm.get_cmap('RdBu')
#Plotting stacked horizontal bar charts for each year to represent
intervals of PM2.5 levels
dfpivot2019.loc[:,['<60','between 61 and 120','between 121 and 180',
'greater than 180']].plot.barh(stacked=True, cmap=cmp,figsize=(15,5),ax=ax1)
dfpivot2018.loc[:,['<60','between 61 and 120','between 121 and
180','greater than 180']].plot.barh(stacked=True, cmap=cmp,
figsize=(15,5),ax=ax2)
dfpivot2017.loc[:,['<60','between 61 and 120','between 121 and
180','greater than 180']].plot.barh(stacked=True, cmap=cmp,
figsize=(15,5),ax=ax3)
dfpivot2016.loc[:,['<60','between 61 and 120','between 121 and
180','greater than 180']].plot.barh(stacked=True, cmap=cmp,
figsize=(15,5),ax=ax4)
#Setting the properties - legend, yaxis and title
ax1.legend().set_visible(False)
ax2.legend().set_visible(False)
ax3.legend().set_visible(False)
```

```
ax4.legend(loc='center left',bbox_to_anchor=(1,0.5))
ax1.get_yaxis().set_visible(False)
ax2.get_yaxis().set_visible(False)
ax3.get_yaxis().set_visible(False)
ax4.get_yaxis().set_visible(False)
ax1.set_title('2019')
ax2.set_title('2018')
ax3.set_title('2017')
ax4.set_title('2016')
```

Output:

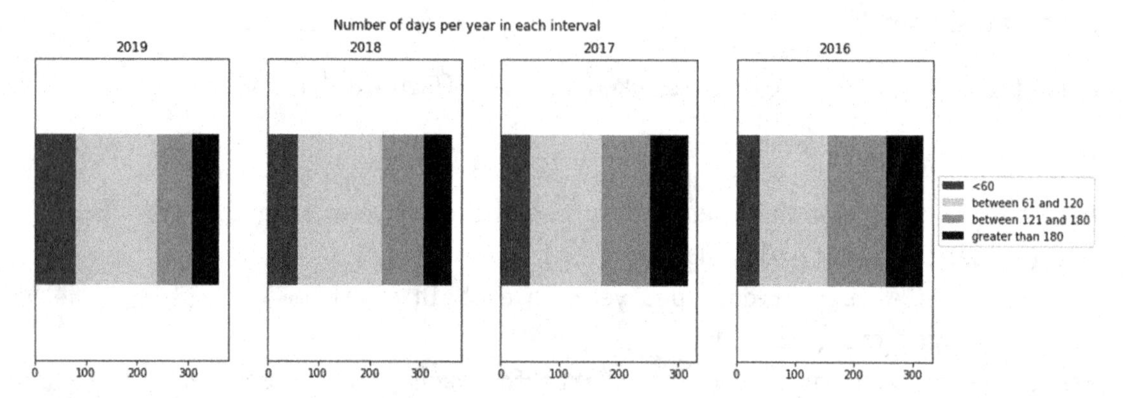

Figure 8-5. *Number of days per year in each interval level for $PM_{2.5}$*

Observation:

It is seen that $PM_{2.5}$ values above 180 μg/m^3 are observed every year, and therefore, to start with, restrictions on major polluting activities, including traffic, could be confined to this category.

NO_2 interval-wise plotting

Likewise, for NO_2, the number of days in each year on which monitored values exceed the 24-hourly standards of 80 μg/m^3 is plotted (Figure 8-6).

First, we create a data frame for NO_2 that captures the number of days in each year with values higher than 80 μg/m^3, as shown in the following.

CODE:

```
#Calculating the number of days in each year with regard to critical days
of NO2 concentration
a=aq2015[(aq2015['NO2']>=80)]['NO2'].count()
b=aq2016[(aq2016['NO2']>=80)]['NO2'].count()
c=aq2017[(aq2017['NO2']>=80)]['NO2'].count()
d=aq2018[(aq2018['NO2']>=80)]['NO2'].count()
e=aq2019[(aq2019['NO2']>=80)]['NO2'].count()
dfno=pd.DataFrame({'years':['2015','2016','2017','2018','2019'],'number of
days with NO2>80 µg':[a,b,c,d,e]})
ax=dfno.plot(kind='bar',figsize=(10,5))
ax.set_xticklabels(['2015','2016','2017','2018','2019'])
ax.set_title("NO2 number of days in each year with critical levels of
concentration")
for p in ax.patches:
    ax.annotate(p.get_height(), (p.get_x() + p.get_width() / 2, p.get_
    height()), ha = 'center', va = 'bottom')
```

Output:

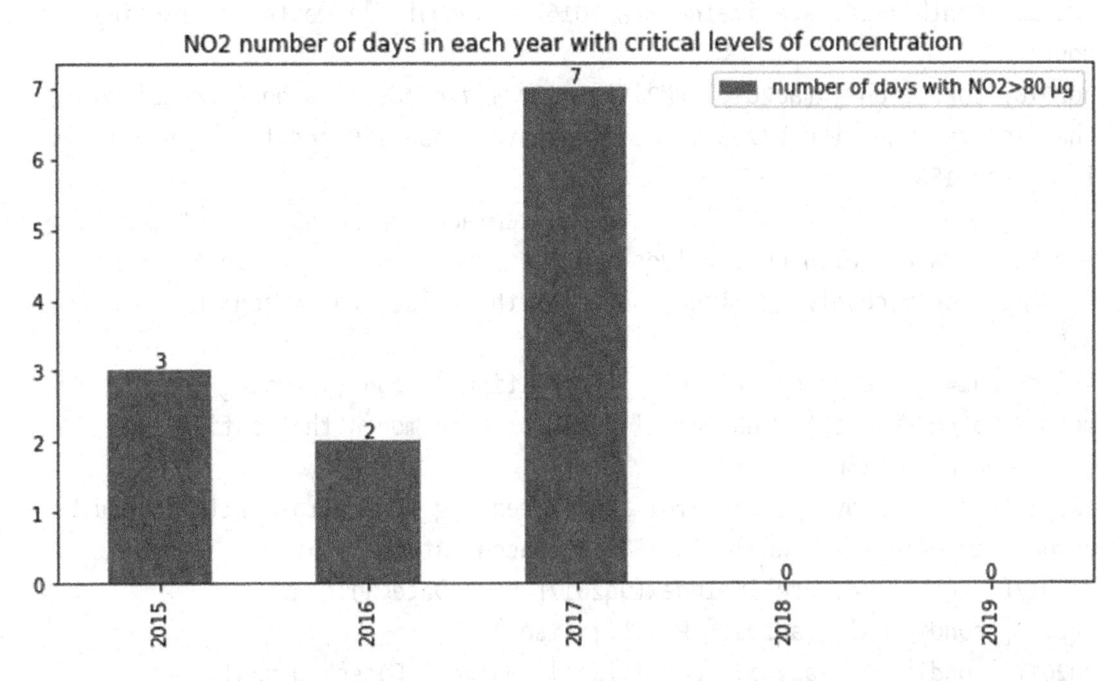

***Figure 8-6.** Number of days per year with critical levels of NO_2 concentration*

Inference: Observed 24-hourly NO_2 values are exceeded only for three of the five years.

Since observed 24-hourly NO_2 values exceed standard only marginally and that too only for a few days, the next step is confined to further analysis of $PM_{2.5}$.

Part 3 of air quality analysis: Identifying the months where $PM_{2.5}$ daily values exceed critical levels on the majority of the days

Before imposing restrictions on activities like vehicular traffic and construction, which significantly contribute to ambient $PM_{2.5}$ concentrations, it is necessary to provide sufficient notice to avoid inconvenience to the general public. Hence, for daily $PM_{2.5}$ values significantly above 180 μg/m^3, we plot temporal variation year-wise during each month of the year. To do this, for each of the twelve months, we capture the number of critical air pollution days every year with 24-hourly $PM_{2.5}$ values exceeding 180 μg/m^3.

First, we create data frames for each year with the number of days in each month where the $PM_{2.5}$ values exceed 180 μg/m^3, as shown in the following.

CODE:

```
#Creating a dataframe for 2016 with the number of days in each month where
the PM2.5 concentration is >180
aq2016['Month']=pd.DatetimeIndex(aq2016['From Date']).month #extracting the
month
aq2016['condition']=(aq2016['PM2.5']>=180 ) # creating a boolean column
that is True when the PM2.5 value is greater than 180 and false when it is
less than 180
aq2016['condition']=aq2016['condition'].replace({False:np.nan}) # replacing
the False values with null values, so that the count method in the next
statement only counts the True values or the values corresponding to PM
2.5>180
selection1=aq2016.groupby('Month')['condition'].count() #Using the groupby
method to calculate the number of days for each month that satisfy the
condition(PM2.5>180)
#Repeating the above process for 2017, creating a dataframe with the number
of days in each month where the PM2.5 concentration is >180
aq2017['Month']=pd.DatetimeIndex(aq2017['From Date']).month
aq2017['condition']=(aq2017['PM2.5']>=180 )
aq2017['condition']=aq2017['condition'].replace({False:np.nan})
```

```
selection2=aq2017.groupby('Month')['condition'].count()
#Repeating the above process for 2018, creating a dataframe with the number
of days in each month where the PM2.5 concentration is >180
aq2018['Month']=pd.DatetimeIndex(aq2018['From Date']).month
aq2018['condition']=(aq2018['PM2.5']>=180 )
aq2018['condition']=aq2018['condition'].replace({False:np.nan})
selection3=aq2018.groupby('Month')['condition'].count()
#Repeating the above process for 2019, creating a dataframe with the number
of days in each month where the PM2.5 concentration is >180
aq2019['Month']=pd.DatetimeIndex(aq2019['From Date']).month
aq2019['condition']=(aq2019['PM2.5']>=180 )
aq2019['condition']=aq2019['condition'].replace({False:np.nan})
selection4=aq2019.groupby('Month')['condition'].count()
```

Now, we concatenate all the DataFrame objects into one object (which we will call 'selectionc') to get a consolidated picture of the number of days in each month where $PM_{2.5} > 180\ \mu g/m^3$, as shown in the following.

CODE:

```
#selectionc data frame is a consolidated dataframe showing month-wise
critical values of PM2.5 for every year
selectionc=pd.concat([selection1,selection1,selection3,selection4],axis=1)
#renaming the columns
selectionc.columns=['2016','2017','2018','2019']
selectionc
```

Output:

Month	2016	2017	2018	2019
1	20	20	23	14
2	3	3	5	3
3	1	1	0	0
4	3	3	0	1
5	3	3	0	2
6	7	7	4	1
7	2	2	0	0
8	0	0	0	0
9	2	2	0	0
10	5	5	5	4
11	13	13	7	11
12	4	4	11	18

We can observe from this table that month 1 (January), month 11 (November), and month 12 (December), are the most critical months for all four years, as these months had the highest number of days with $PM_{2.5}$ > 180 μg/m³.

Now that we have all the data in place, let us visualize the critical days for $PM_{2.5}$ (Figure 8-7), using the following code.

CODE:

```
#creating a bar chart representing number of days with critical levels of
PM2.5(>180) concentrations
ax=selectionc.plot(kind='bar',figsize=(20,7),width=0.7,align='center',color
map='Paired')
bars = ax.patches
#creating patterns to represent each year
patterns =('-','x','/','O')
#ax.legend(loc='upper left', borderpad=1.5, labelspacing=1.5)
ax.legend((patterns),('2016','2017','2018','2019'))
hatches = [p for p in patterns for i in range(len(selectionc))]
#setting a pattern for each bar
```

```
for bar, hatch in zip(bars, hatches):
    bar.set_hatch(hatch)
#Labelling the months, the X axis and Y axis
ax.set_xticklabels(['Jan','Feb','Mar','Apr','May','June','July','Aug','Sept
','Oct','Nov','Dec'],rotation=30)
ax.set_xlabel('Month',fontsize=12)
ax.set_ylabel('Number of days with critical levels of PM2.5',fontsize=12)
#Labelling the bars
for i in ax.patches:
    ax.text(i.get_x()-.003, i.get_height()+.3,
            round(i.get_height(),2), fontsize=10,
                color='dimgrey')
ax.legend()
ax.set_title("Number of days with critical levels of PM2.5 in each month of
years 2016-19")
```

Output:

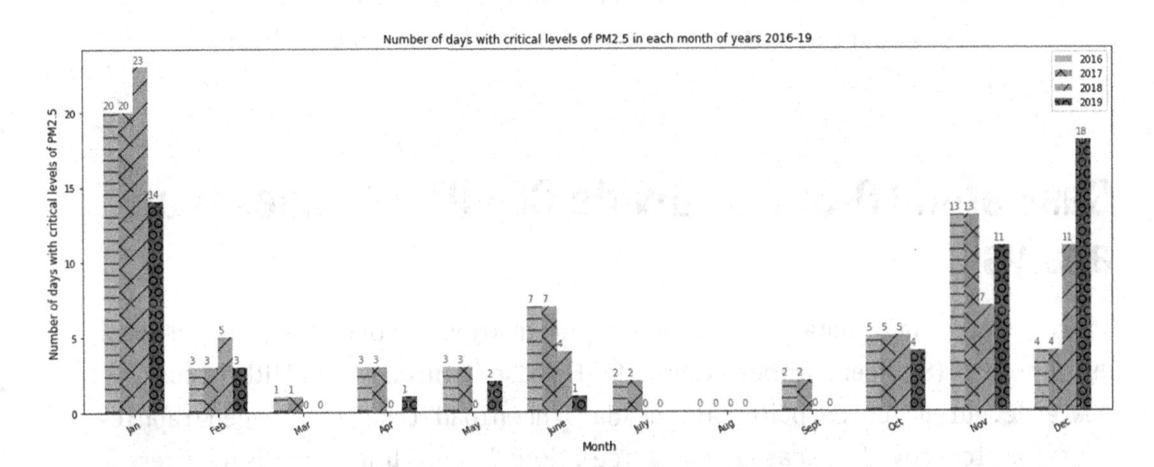

***Figure 8-7.** $PM_{2.5}$ – Number of days with critical levels per month per year*

Step 4: Drawing inferences based on analysis and visualizations

From the preceding graph, it is observed that most of the critically polluted days fall in January, November, and December. Therefore, based on daily average concentrations of $PM_{2.5}$ recorded over the past four years, restrictions on vehicular traffic, construction activities, use of diesel pump sets, diversion of traffic entering Delhi from neighboring states, and other similar activities are likely to be imposed in January, November, and December. To make such decisions for Delhi as a whole, analysis of data from other monitoring stations would also be necessary. Dissemination of data and analysis on the preceding lines would help people prepare in advance for restrictions and also appreciate the rationale behind such measures.

The approach demonstrated in the preceding, using data analysis as a tool to assist in air quality management, uses the data recorded at one monitoring station located at the Netaji Subhas Institute of Technology (NSIT), Delhi. The methodology could be carried forward on the following lines.

a. Repeat the preceding step for NO_2 to show critical months that account for most of the days with NO_2 recorded values exceeding 24-hourly standards. Doing this exercise would again help identify months facing pollution levels of concern for both parameters, $PM_{2.5}$ and NO_2, and plan.

b. Repeat the analysis carried out with data from the air quality monitoring station at NSIT with the use of similar data from other stations so that interventions for Delhi as a whole could be planned.

Case study 8-3: Worldwide COVID-19 cases – an analysis

This dataset contains data about the geographic distribution of COVID-19 cases as of 29th June 2020 (Source: European Center for Disease Control, source URL: `https://www.ecdc.europa.eu/en/publications-data/download-todays-data-geographic-distribution-covid-19-cases-worldwide`). Note that this link contains the latest data, but we have used the data as on 29th June (the link to the dataset is provided in the "Technical requirements" section at the beginning of the chapter).

Questions that we want to answer through our analysis include:

1. Which are the countries with the worst mortality rates, maximum cases, and the most fatalities?

2. What is the monthly trend vis-à-vis the number of cases and fatalities since the start of the pandemic?

3. In some of the countries, lockdowns were imposed to help flatten the curve. Did this measure aid in reducing the number of cases?

Step 1: Importing the data and examining the characteristics of the dataset

Read the dataset and examine the first five rows (using the *head* method) using the *pd.read_excel* function:

CODE:

```
df=pd.read_excel('COVID-19-geographic-distribution-worldwide-2020-06-29.xlsx')
df.head()
```

Output:

	dateRep	day	month	year	cases	deaths	countriesAndTerritories	geoId	countryterritoryCode	popData2019	continentExp
0	2020-06-29	29	6	2020	351	18	Afghanistan	AF	AFG	38041757.0	Asia
1	2020-06-28	28	6	2020	165	20	Afghanistan	AF	AFG	38041757.0	Asia
2	2020-06-27	27	6	2020	276	8	Afghanistan	AF	AFG	38041757.0	Asia
3	2020-06-26	26	6	2020	460	36	Afghanistan	AF	AFG	38041757.0	Asia
4	2020-06-25	25	6	2020	234	21	Afghanistan	AF	AFG	38041757.0	Asia

Get information about the data type of each column and the number of non-null values in each column (using the *info* method).

CODE:

```
df.info()
```

Output:

```
<class 'pandas.core.frame.DataFrame'>
RangeIndex: 26562 entries, 0 to 26561
Data columns (total 11 columns):
dateRep                     26562 non-null datetime64[ns]
day                         26562 non-null int64
month                       26562 non-null int64
year                        26562 non-null int64
cases                       26562 non-null int64
```

```
deaths                      26562 non-null int64
countriesAndTerritories     26562 non-null object
geoId                       26455 non-null object
countryterritoryCode        26498 non-null object
popData2019                 26498 non-null float64
continentExp                26562 non-null object
dtypes: datetime64[ns](1), float64(1), int64(5), object(4)
memory usage: 2.2+ MB
```

Get summary statistics for each column (using the *describe* method) and obtain the values of the count, min, max, standard deviation, and percentiles:

CODE:

```
df.describe()
```

Output:

	day	month	year	cases	deaths	popData2019
count	26562.000000	26562.000000	26562.000000	26562.000000	26562.000000	2.649800e+04
mean	16.207929	4.194790	2019.997478	380.722611	18.882690	4.689196e+07
std	8.745421	1.555569	0.050161	2172.430663	121.386696	1.675462e+08
min	1.000000	1.000000	2019.000000	-2461.000000	-1918.000000	8.150000e+02
25%	9.000000	3.000000	2020.000000	0.000000	0.000000	1.919968e+06
50%	17.000000	4.000000	2020.000000	4.000000	0.000000	8.776119e+06
75%	24.000000	5.000000	2020.000000	68.000000	1.000000	3.194979e+07
max	31.000000	12.000000	2020.000000	54771.000000	4928.000000	1.433784e+09

Step 2: Data wrangling

In this step, we will:

- Check if the data types of the columns are accurately identified. If not, change the data types: From the output of the *info* method, we see that all data types of the columns have been correctly identified.

- Rename the columns if necessary: In the following code, we are renaming the columns of the DataFrame.

CODE:

```
#changing the column names
df.columns=['date','day','month','year','cases','deaths','country',
'old_country_code','country_code','population','continent']
```

- Drop any unnecessary columns or rows:
- We see the country code column is present twice (with two different names: 'old_country_code' and 'country_code') in the DataFrame, hence we remove one of the columns ("old_country_code"):

  CODE:

```
#Dropping the redundant column name
df.drop(['old_country_code'],axis=1,inplace=True)
```

- Remove any extraneous data that does not add any value:

  There are no blank spaces, special characters, or any other extraneous characters in this DataFrame. We see that there is data for only one day in December 2019; hence we remove data for this month and create a new DataFrame (df1) for the remaining 11 months.

  CODE:

```
df1=df[df.month!=12]
```

- Check if there are any null values, using the *isna* or *isnull* method, and apply appropriate methods to deal with them if they are present:

  Calculating the percentage of null values:

  CODE:

```
df1.isna().sum().sum()/len(df1)
```

  Output:

```
0.008794112096622004
```

  Since the percentage of null values is less than 1%, we drop the null values in the following step.

CODE:

```
df1.dropna(inplace=True)
```

- Aggregate the data if the data is in a disaggregated format:

  The data in this DataFrame is not in an aggregated format, and we convert it into this format using the *groupby* method in this step. We can group either by country, by continent, or by date. Let us group by the name of the country.

  CODE:

```
#Aggregating the data by country name
df_by_country=df1.groupby('country')['cases','deaths'].sum()
df_by_country
```

  Output (only first nine rows shown):

	cases	deaths
country		
Afghanistan	30967	721
Albania	2402	55
Algeria	13273	897
Andorra	855	52
Angola	267	11
Anguilla	3	0
Antigua_and_Barbuda	65	3
Argentina	57731	1217
Armenia	25127	433

The preceding output shows a consolidated picture of the number of cases and deaths for each country.

Let us add another column to this aggregated DataFrame - the mortality rate, which is the ratio of the number of deaths to the number of cases.

CODE:

```
#Adding a new column for the mortality rate which is the ratio of the
number of deaths to cases
df_by_country['mortality_rate']=df_by_country['deaths']/df_by_
country['cases']
```

Step 3: Visualizing the data

In our first visualization in this case study, we use the aggregated data in the DataFrame, "df_by_country", to display the top twenty countries by mortality rate (Figure 8-8).

CODE:

```
#Sorting the values for the mortality rate in the descending order
plt.figure(figsize=(15,10))
ax=df_by_country['mortality_rate'].sort_values(ascending=False).head(20).
plot(kind='bar')
ax.set_xticklabels(ax.get_xticklabels(), rotation=45, ha="right")
for p in ax.patches:
    ax.annotate(p.get_height().round(2),(p.get_x()+p.get_width()/2,p.get_he
ight()),ha='center',va='bottom')
ax.set_xlabel("Country")
ax.set_ylabel("Mortality rate")
ax.set_title("Countries with highest mortality rates")
```

Output:

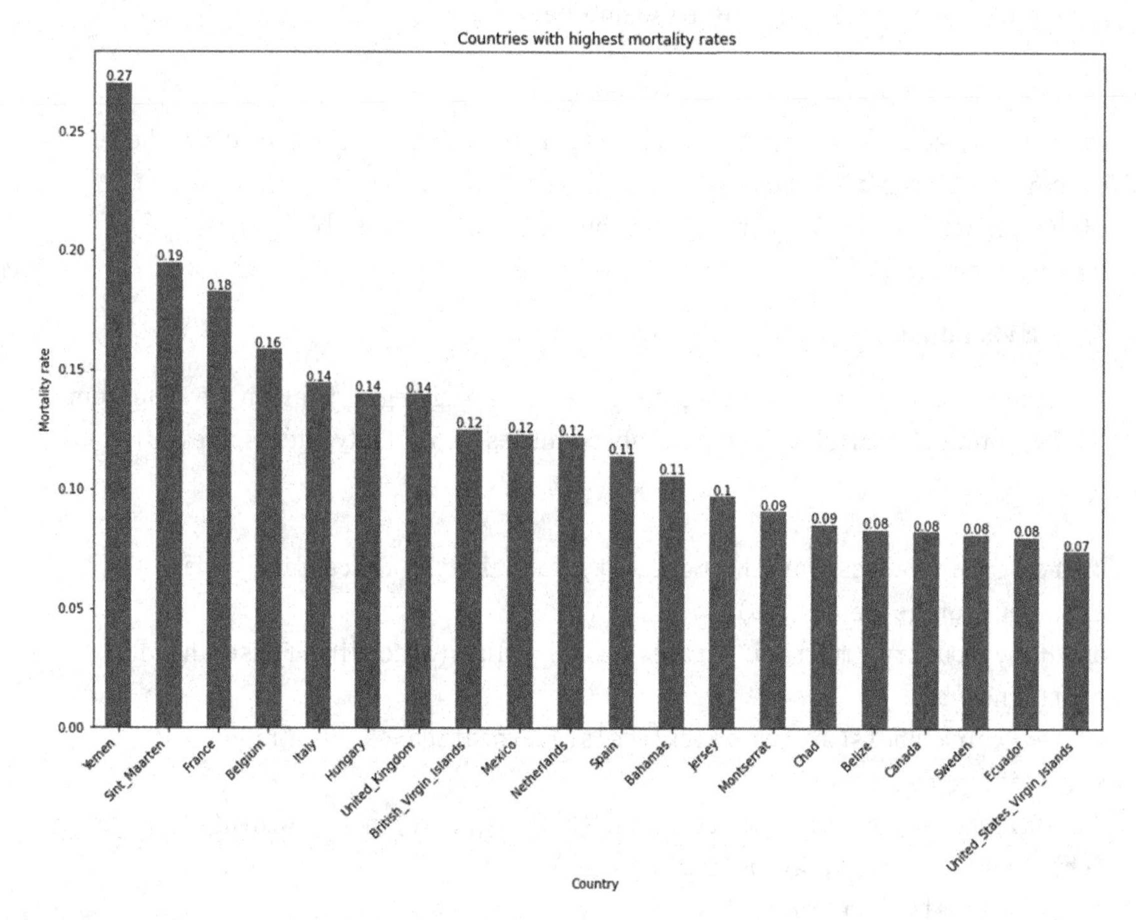

Figure 8-8. *Bar chart depicting countries with the highest mortality rates for COVID-19*

In the second visualization, we display the ten countries with the highest number of COVID-19 cases, using a pie chart, as shown in Figure 8-9.

CODE:

```
#Pie chart showing the countries with the highest number of COVID cases
df_cases=df_by_country['cases'].sort_values(ascending=False)
ax=df_cases.head(10).plot(kind='pie',autopct='%.2f%%',labels=df_cases.
index,figsize=(12,8))
ax.set_title("Top ten countries by case load")
```

Output:

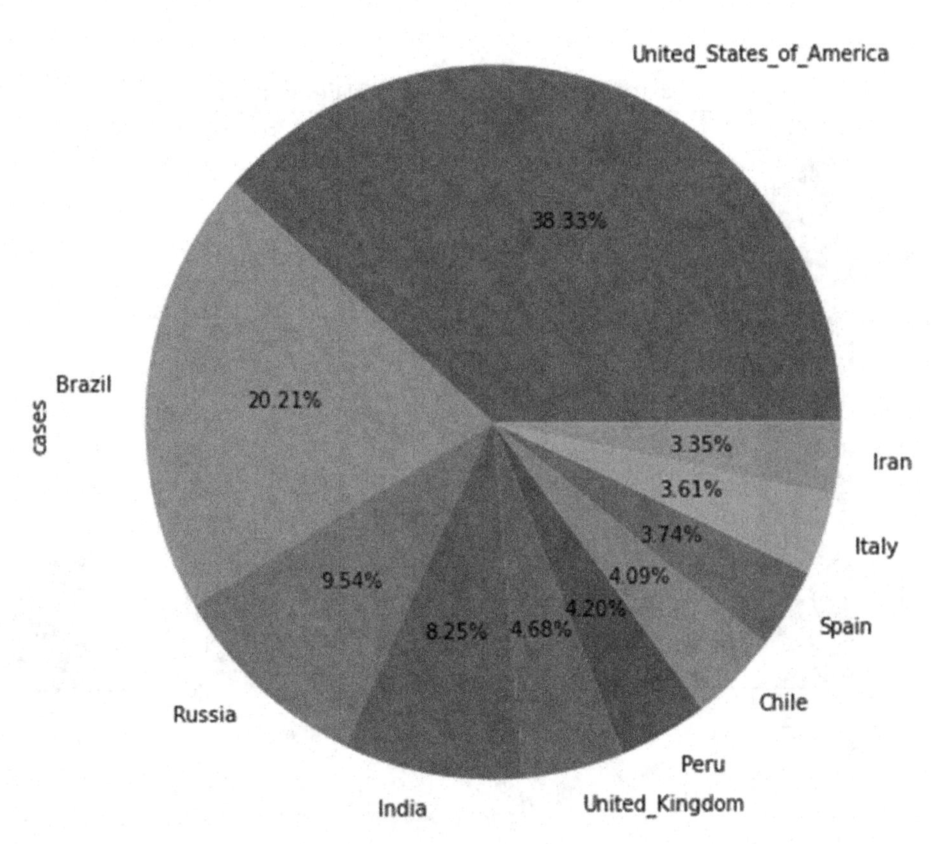

Figure 8-9. *Pie chart depicting the share of the top ten countries by COVID-19 cases*

In the next visualization, we find out the five countries that have suffered the most in terms of loss to human life from the COVID-19 pandemic, using a bar chart (Figure 8-10).

CODE:

```
#sorting the number of deaths in the descending order
plt.figure(figsize=(10,6))
ax=df_by_country['deaths'].sort_values(ascending=False).head(5).
plot(kind='bar')
ax.set_xticklabels(ax.get_xticklabels(), rotation=45, ha="right")
for p in ax.patches:
    ax.annotate(p.get_height(),(p.get_x()+p.get_width()/2,p.get_height()),
    ha='center',va='bottom')
```

```
ax.set_title("Countries suffering the most fatalities from COVID-19")
ax.set_xlabel("Countries")
ax.set_ylabel("Number of deaths")
```

Output:

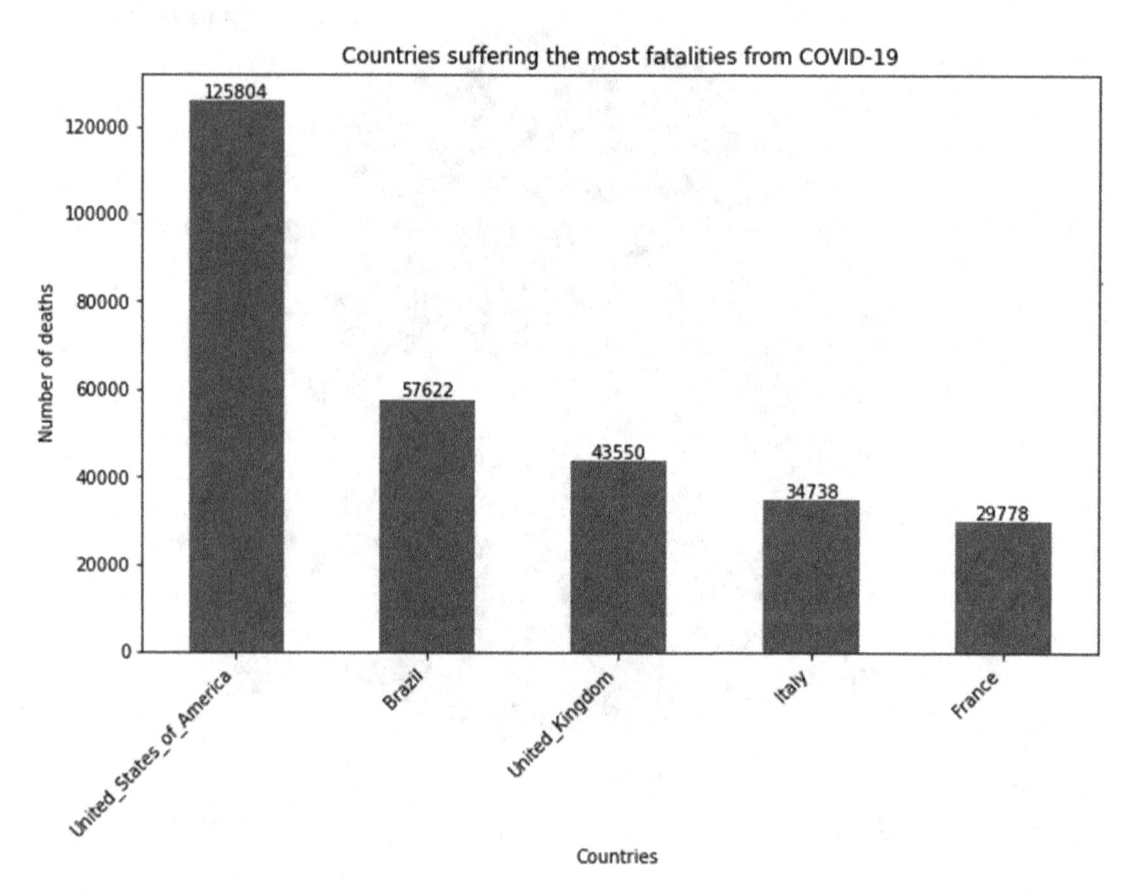

Figure 8-10. *Bar chart depicting the five countries with the maximum fatalities*

Now, we plot line graphs to see the month-wise trend in the number of COVID-19 cases and fatalities.

To plot the line graphs, we first aggregate the data by month and then plot two line graphs side by side, showing the number of cases and deaths, as shown in Figure 8-11.

CODE:

```
df_by_month=df1.groupby('month')['cases','deaths'].sum()
fig=plt.figure(figsize=(15,10))
ax1=fig.add_subplot(1,2,1)
```

```
ax2=fig.add_subplot(1,2,2)
df_by_month['cases'].plot(kind='line',ax=ax1)
ax1.set_title("Total COVID-19 cases across months in 2020")
ax1.set_xlabel("Months in 2020")
ax1.set_ylabel("Number of cases(in million)")
df_by_month['deaths'].plot(kind='line',ax=ax2)
ax2.set_title("Total COVID-19 deaths across months in 2020")
ax2.set_xlabel("Months in 2020")
ax2.set_ylabel("Number of deaths")
```

Output:

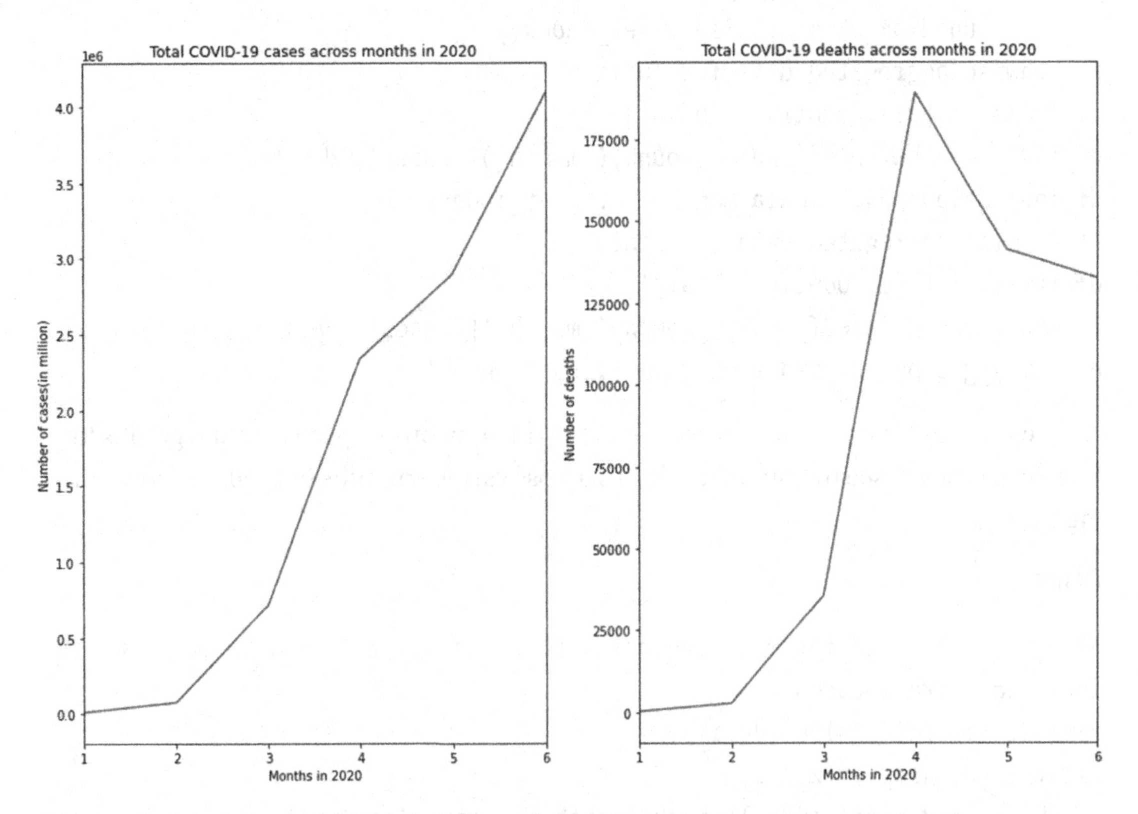

***Figure 8-11.** Impact of lockdown on flattening the curve*

Many countries imposed a lockdown to stem the tide of increasing cases and flatten the curve. We now look at four countries – India, the UK, Italy, and Germany – where lockdowns were imposed in March, to see if this measure had the desired impact.

First, we create DataFrame objects for each of these countries, with data aggregated month-wise.

CODE:

```
#Creating DataFrames for each country
#Monthwise aggregated data for Germany
df_germany=df1[df1.country=='Germany']
df_germany_monthwise=df_germany.groupby('month')['cases','deaths'].sum()
df_germany_grouped=df_germany_monthwise.reset_index()
#Monthwise aggregated data for UK
df_uk=df1[df1.country=='United_Kingdom']
df_uk_monthwise=df_uk.groupby('month')['cases','deaths'].sum()
df_uk_grouped=df_uk_monthwise.reset_index()
#Monthwise aggregated data for India
df_india=df1[df1.country=='India']
df_india_monthwise=df_india.groupby('month')['cases','deaths'].sum()
df_india_grouped=df_india_monthwise.reset_index()
#Monthwise aggregated data for Italy
df_italy=df1[df1.country=='Italy']
df_italy_monthwise=df_italy.groupby('month')['cases','deaths'].sum()
df_italy_grouped=df_italy_monthwise.reset_index()
```

Now, we use the DataFrame objects created in the previous steps to plot line graphs for these countries to see the number of cases across various months in 2020, as shown in Figure 8-12.

CODE:

```
#Plotting the data for four countries (UK, India, Italy and Germany) where
lockdowns were imposed
fig=plt.figure(figsize=(20,15))
ax1=fig.add_subplot(2,2,1)
df_uk_grouped.plot(kind='line',x='month',y='cases',ax=ax1)
ax1.set_title("Cases in UK across months")
ax2=fig.add_subplot(2,2,2)
df_india_grouped.plot(kind='line',x='month',y='cases',ax=ax2)
ax2.set_title("Cases in India across months")
ax3=fig.add_subplot(2,2,3)
df_italy_grouped.plot(kind='line',x='month',y='cases',ax=ax3)
```

```
ax3.set_title("Cases in Italy across months")
ax4=fig.add_subplot(2,2,4)
df_germany_grouped.plot(kind='line',x='month',y='cases',ax=ax4)
ax4.set_title("Cases in Germany across months")
```

Output:

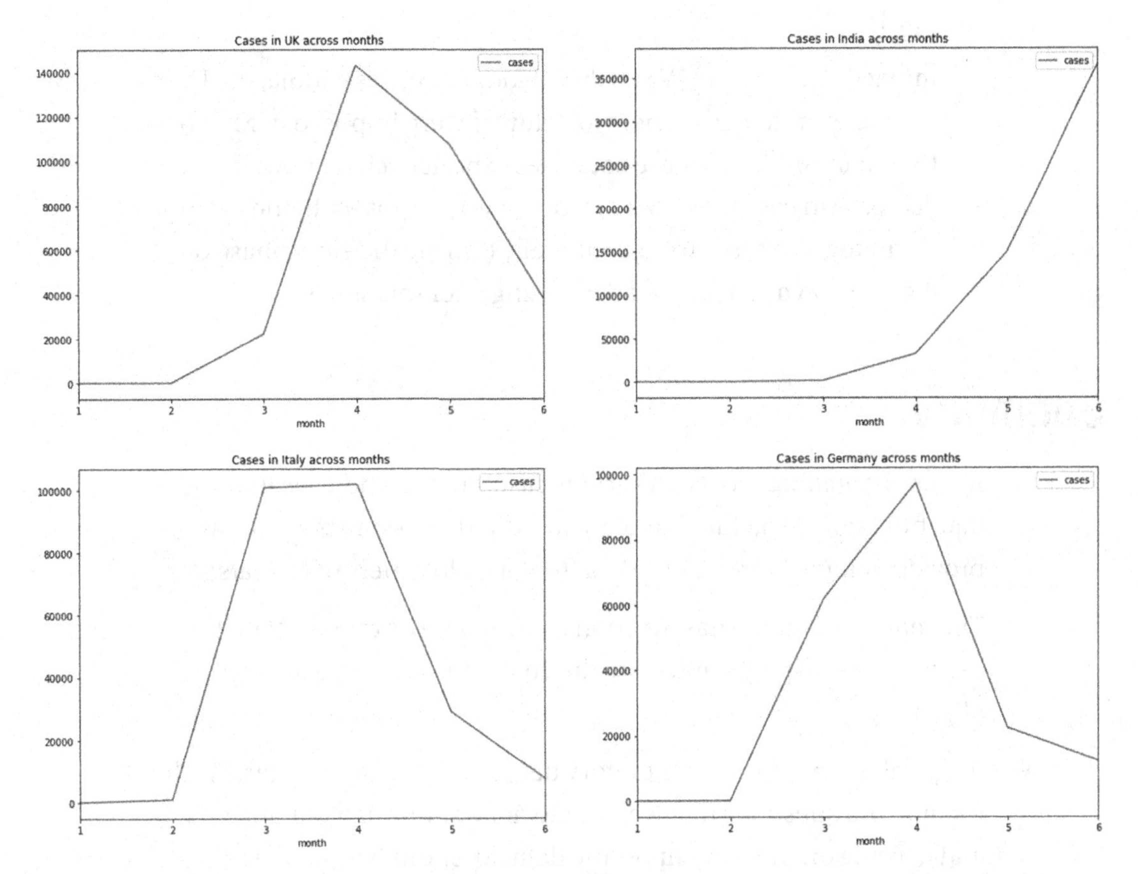

Figure 8-12. *Total cases in UK, India, Germany, and Italy in the first 6 months of 2020*

Step 4: Drawing inferences based on analysis and visualizations

- Number of cases: The United States, Brazil, Russia, India, and the UK had the highest number of cases.
- Number of deaths: The United States, Brazil, the UK, Italy, and France had the highest death tolls.

- Mortality rate: Yemen, St. Maarten, France, Belgium, and Italy had the highest mortality rates.
- Trends:
    - The total number of cases has been increasing steadily, while the total number of fatalities (deaths) has shown a decrease after April.
    - Impact of lockdown: We analyzed four countries - India, the UK, Germany, and Italy - where lockdowns were imposed in March. Except for India, all these countries experienced an overall decrease in cases after the lockdown was imposed. In the UK and Germany, the cases went up initially (during the early phase of the lockdown) but started decreasing after this spike.

Summary

- In this chapter, we looked at various case studies where we imported data from both structured and unstructured data sources. Pandas provides support for reading data from a wide variety of formats.
- The requests module has functions that enable us to send HTTP requests to web pages and store the content from the page in an object.
- A typical descriptive or exploratory data analysis of a case starts with framing the questions that we want to answer through our analysis and figuring out how to import the data. After this, we get more information about the data - the meanings of various columns, the units of measurement, the number of missing values, the data types of different columns, and so on.
- Data wrangling, where we prepare, clean, and structure the data to make it suitable for analysis, is the crux of descriptive or exploratory data analysis. Typical activities involved removing extraneous data, handling null values, renaming columns, aggregating data, and changing data types.

- Once the data is prepared and made suitable for analysis, we visualize our data using libraries like Matplotlib, Seaborn, and Pandas to help us gain insights that would answer the questions we initially framed.

Review Exercises

Question 1 (mini case study)

Consider the first table on the following web page: `https://en.wikipedia.org/wiki/Climate_of_South_Africa`. It contains data about the maximum and minimum temperatures (in degrees centigrade) in various cities in South Africa, during summers and winters.

- Use the appropriate method from the *requests* module to send a *get* request to this URL and store the data from the first table on this page in a Pandas DataFrame.
- Rename the columns as follows: 'City', 'Summer(max)', 'Summer(min)', 'Winter(max)', 'Winter(min)'.
- Replace the negative value in the first row of the 'Winter(min)' column with 0, and convert the data type of this column to *int64*.
- Plot a graph to display the hottest cities in South Africa during summers (use the Summer(max) column).
- Plot a graph to display the coldest cities in South Africa during the winters (use the Winter(min) column).

Question 2

The weekly wages of ten employees (with the initials A–J) are as follows: 100, 120, 80, 155, 222, 400, 199, 403, 345, 290. Store the weekly wages in a DataFrame.

- Plot a bar graph to display the wages in the descending order
- Label each of the bars in the bar graphs using the *annotate* method

Question 3

1. Module for sending HTTP requests	a. URL of a web page
2. *Get* method	b. req.text
3. Argument passed to *get* method	c. Fetches information using a given URL
4. Attribute containing Unicode content	d. Requests

Question 4

The *read_html* Pandas function reads

1. All the HTML content on the web page
2. HTML tags in a web page
3. All the HTML tables as a list of DataFrame objects
4. HTML lists in a web page

Answers

Question 1

CODE:

```
import requests
import pandas as pd
import seaborn as sns
import matplotlib.pyplot as plt
%matplotlib inline
url='https://en.wikipedia.org/wiki/Climate_of_South_Africa'
#making a get request to the URL
req = requests.get(url)
#storing the HTML data in a DataFrame
data = pd.read_html(req.text)
#Reading the first table
df=data[0]
#Renaming the columns
df.columns=['City','Summer(max)','Summer(min)','Winter(max)','Winter(min)']
#Replacing the negative value with 0
df['Winter(min)']=df['Winter(min)'].str.replace(r"-2","0")
```

```
#Changing the data type from object to int64
df['Winter(min)']=df['Winter(min)'].astype('int64',errors='ignore')
#Using the city as the index to facilitate plotting
df1=df.set_index('City')
#Hottest five cities during Summer
df1['Summer(max)'].sort_values(ascending=False).head(5).plot(kind='bar')
#Coldest five cities during Winter
df1['Winter(min)'].sort_values().head(5).plot(kind='bar')
```

Question 2

CODE:

```
numbers=pd.Series([100,120,80,155,222,400,199,403,345,290])
#converting the data to a DataFrame
numbers.to_frame()
#labelling the index
numbers.index=list('ABCDEFGHIJ')
#labelling the column
numbers.columns=['Wages']
ax=numbers.sort_values(ascending=False).plot(kind='bar')
#labelling the bars
for p in ax.patches:
    ax.annotate(p.get_height(),(p.get_x()+p.get_width()/2,p.get_height()),h
a='center',va='bottom')
```

Question 3

1-d; 2-c; 3-a; 4-b

Question 4

Option 3

Content from each table on the web page is stored in a separate DataFrame object.

CHAPTER 9

Statistics and Probability with Python

In the previous chapter, we learned about how to apply your knowledge of data analysis by solving some case studies.

Now, in the final part of this book, we learn about essential concepts in statistics and probability and understand how to solve statistical problems with Python. The topics that we cover include permutations and combinations, probability, rules of probability and Bayes theorem, probability distributions, measures of central tendency, dispersion, skewness and kurtosis, sampling, central limit theorem, and hypothesis testing. We also look at confidence levels, level of significance, p-value, hypothesis testing, parametric tests (one- and two-sample z-tests, one- and two-sample t-tests, paired tests, analysis of variance [ANOVA]), and nonparametric tests (chi-square test).

Permutations and combinations

Let us look at a few definitions, formulae, and examples that will help us understand the concepts of permutations and combinations.

Combinations: The various ways in which we can select a group of objects.

The following formula gives the number of combinations we can form from a given number of objects:

$$nc_r = \frac{n!}{r!(n-r)!}$$

G. Rajagopalan, *A Python Data Analyst's Toolkit*, https://doi.org/10.1007/978-1-4842-6399-0_9

In the preceding formula, *n* is the total number of objects in the set from which a smaller subset of objects is drawn, *c* is the number of combinations, and *r* is the number of objects in the subset. The exclamation mark symbol (!) denotes the factorial of a number. For example, x! is the product of all integers from 1 up to and including x.

$$(x! = x*(x-1)*(x-2) .... *1)$$

Let us now solve a simple question involving combinations.

Question: Find the number of ways in which an ice cream sundae containing three flavors can be created out of a total of five flavors.

Answer: Let the five flavors be A, B, C, D, and E. Working out this problem manually, the following combinations can be obtained:

A, B, C|B, C, D|A, C, D|A, B, D|C, D, E|B, D, E|A, B, E|A, D, E|A, C, E|B, C, E

There are ten combinations, as we can see. If we apply the nc_r formula, where *n* is 5 and *r* is 3, we get the same answer ($5C_3 = 10$).

Let us now look at what permutations are.

Permutations are similar to combinations, but here, the order in which the objects are arranged matters.

The following formula gives the number of permutations:

$$n_{P_r} = \frac{n!}{(n-r)!}$$

Considering the same ice cream example, let us see how many permutations we can obtain, that is, the number of ways in which three flavors can be selected and arranged out of a total of five flavors.

1. ABC|CBA|BCA|ACB|CAB|BAC
2. BCD|CDB|BDC|CBD|DBC|DCB
3. ACD|ADC|DAC|DCA|CAD|CDA
4. ABD|ADB|BAD|BDA|DAB|DBA
5. CDE|CED|DCE|DEC|ECD|EDC

6. BDE|BED|DBE|DEB|EBD|EDB
7. ABE|AEB|BEA|BAE|EAB|EBA
8. ADE|AED|DAE|DEA|EAD|EDA
9. ACE|AEC|CAE|CEA|EAC|ECA
10. BCE|BEC|CBE|CEB|EBC|ECB

As we can see, we can get six possible arrangements for each combination. Hence, the total number of arrangements = 10*6 = 60. The formula n_{P_r} (where n=5 and r=3) also gives the same answer (60).

Another approach to solving questions involving permutations is as follows:

1. First, select the items: Select three flavors from five in $5C_3$ ways
2. Now arrange the three items in 3! ways
3. Multiply the results obtained in step 1 and step 2. Total number of permutations = $5C_3 * 3! = 60$

Now that we have understood the concepts of permutations and combinations, let us look at the essentials of probability.

Probability

Given below are a few important concepts related to probability.

Random experiment: This is any procedure that leads to a defined outcome and can be repeated any number of times under the same conditions.

Outcome: The result of a single experiment.

Sample space: Exhaustive list containing all possible outcomes of an experiment.

Event: An event can comprise a single outcome or a combination of outcomes. An event is a subset of the sample space.

Probability: A quantitative measure of the likelihood of the event. The probability of any event always lies between 0 and 1. 0 denotes that the event is not possible, while 1 indicates that the event is certain to occur.

If the letter X denotes our event, then the probability is given by the notation P(X)= N(X)/N(S)

Where N(X)=number of outcomes in event x

N(S)= total number of outcomes in the sample space

Solved example: Probability

The following is a simple probability question.

Question: In an experiment, a die is rolled twice. Find the probability that the numbers obtained in the two throws add up to 10.

Solution:

Event A: The first die is rolled.

Event B: The second die is rolled.

Sample space: A die contains the numbers 1 to 6, which are equally likely to appear. The total number of outcomes, when one die is rolled, is six. Since events "A" and "B" are independent, the total number of outcomes for both the events = 6*6 = 36.

Event X: The sum of the two numbers is 10. The possible outcomes that lead to this result include {4,6}, {6,4}, and {5,5}; that is, three possible outcomes lead to a sum of 10.

P(X)=Probability of obtaining a sum of 10=Number of outcomes in event X/Total Sample Space=3/36=0.0833

Rules of probability

Let us understand the various rules of probability, explained in Table 9-1.

***Table 9-1.** Rules of Probability*

Rule	Description	Formula	Venn diagram
Addition rule	The addition rule determines the probability of either of two events occurring.	P (A U B) = P(A)+P(B)-P (A ∩ B)	AUB A B

(continued)

Table 9-1. *(continued)*

Rule	Description	Formula	Venn diagram
Special rule of addition	This rule applies to mutually exclusive events. Mutually exclusive events are those that cannot co-occur. For mutually exclusive events, the probability of either of the events occurring is simply the sum of the probability of each of the events.	$P(A \cup B) = P(A)+P(B)$	MUTUALLY EXCLUSIVE A B
Multiplication rule	The multiplication rule is a general rule that provides the probability of two events occurring together.	$P(A \cap B)=P(A)*P(B/A)$ P(B/A) is the *conditional probability* of event B happening given that event A has already occurred.	A $A \cap B$ B
Special rule of multiplication	This rule applies to independent events. For independent events, the probability of the events occurring together is simply the product of probabilities of the events.	$P(A \cap B)=P(A)*P(B)$	A $A \cap B$ B

Note that the formulae listed in Table 9-1 provide the rules for two events, but these can be extended to any number of events.

Conditional probability

Conditional probability involves calculating the probability of an event, after taking into consideration the probability of another event that has already occurred. Consider Figure 9-1, which illustrates the principle of conditional probability.

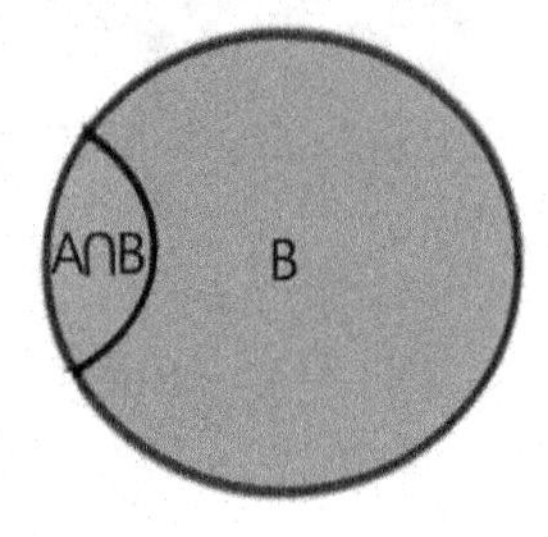

Figure 9-1. *Conditional probability*

Figure 9-1 shows that if the event "A" is dependent on event "B," then the sample space is event "B" itself. For example, let A be the event that a customer purchases a product from an online retailer. Let the probability of the event be 0.5., or in other words, P(A)=0.5.

Now, let B be the event in which the product that the customer intends to purchase has received a negative review. The probability of the customer buying the product may now be less than what it was earlier due to this negative review. Let us say that now there is only a 30% chance that they purchase the product. In other words, P(A/B) = probability of the customer buying a product given that it has received a negative review = 0.3.

The formula for conditional probability is $P(A/B) = P(A \cap B)/P(B)$.

Bayes theorem

Bayes theorem is a theorem used to calculate the conditional probability of an event, given some evidence related to the event. Bayes theorem establishes a mathematical relationship between the probability of an event and prior knowledge of conditions related to it. As evidence related to an event accumulates, the probability of this event can be determined more accurately.

Questions involving Bayes theorem are different from conventional probability questions, where we generally calculate the probability of events that may occur in the future. For instance, in a simple conditional probability question, we might be asked to calculate the probability of a person getting diabetes, given that they are obese. In Bayes theorem, we go backward and calculate the probability of a person being obese, given that they have diabetes. That is, if a person tested positive for diabetes, Bayes theorem tests the hypothesis that he is obese. The formula for Bayes theorem is as follows:

$$P(A/B) = \frac{P(B/A) * P(A)}{P(B)}$$

P(A/B), also known as the posterior probability, is what we want to compute, that is, the probability of the hypothesis being true, given the data on hand.

P(B/A) is the probability of obtaining the evidence, given the hypothesis.

P(A) is the *prior probability*, that is, the probability of the hypothesis being true before we have any data related to it.

P(B) is the general probability of occurrence of the evidence, without any hypothesis, also called the *normalizing constant.*

Applications of Bayes theorem: Given below are a few areas where Bayes theorem can be applied.

- Medical diagnostics
- Finance and risk assessment
- Email spam classification
- Law enforcement and crime detection
- Gambling

Now, let us understand the practical application of Bayes theorem in a few of these areas using a couple of examples.

Application of Bayes theorem in medical diagnostics

Bayes theorem has applications in the field of medical diagnostics, which can be understood with the help of the following hypothetical example.

Question: Consider a scenario where a person has tested positive for an illness. This illness is known to impact about 1.2% of the population at any given time. The diagnostic test is known to have an accuracy of 85%, for people who have the illness. The test accuracy is 97% for people who do not have the illness. What is the probability that this person suffers from this illness, given that they have tested positive for it?

Solution:

Assessing the accuracy of medical tests is one of the applications of Bayes theorem.

Let us first define the events:

A: The person has the illness, also called the hypothesis.

~A: The person does not have the illness.

B: The person has tested positive for the illness, also called the evidence.

P(A/B): Probability that this person has the illness given that they have tested positive for it, also called the *posterior probability* (which is what we need to calculate).

P(B/A): Probability that the person has tested positive for it given that they have the illness. This value is 0.85 (as given in the question that this test has 85% accuracy for people who suffer from the illness).

P(A): *Prior probability* or the probability that the person has the illness, without any evidence (like a medical test). This value is 0.012 (as given in the question that this illness affects 1.2% of the population).

P(B): Probability that this person has tested positive for this illness. This probability can be calculated in the following manner.

There are two ways this person could test positive for this illness:

- They have the illness and have tested positive (true POSITIVE) - the probability of this occurring can be calculated as follows:

  P(B/A)*P(A)=0.85*0.012=0.0102.

- They do not have the illness, but the test was inaccurate, and they have tested positive for it (false positive) – the probability of this occurring can be calculated as follows:

  P(B/~A)*P(~A)=(1-0.97)*(1-0.012)=0.02964.

Here, P(B/~A) refers to the probability that the test was positive for a person who did not have the illness, that is, the probability that the test was inaccurate for the person who does not suffer from the illness.

Since this test is 97% accurate for people who do not have this illness, it is inaccurate for 3% of the cases.

In other words, P(B/~A) = 1-0.97.

Similarly, P(~A) refers to the probability that the person does not have the illness. Since we have been provided the data that the incidence of this illness is 1.2%, P(~A) is = 1-0.012.

P(B), the denominator in the Bayes theorem equation, is the union of the preceding two probabilities = (P(B/A)*P(A)) + (P(B/~A)*P(~A))=0.0102+0.2964=0.03984

We can now calculate our final answer by plugging in the values for the numerator and denominator in the Bayes theorem formula.

P(A/B)=P(B/A)*P(A)/P(B) =0.85*0.012 / 0.03984 = 0.256

Using the Bayes theorem, we can now conclude that even with a positive medical test, this person only has a 25.6% chance of suffering from this illness.

Another application of Bayes theorem: Email spam classification

Let us look at another application of Bayes theorem in the area of email spam classification. Before the era of spam filters, there was no way of separating unsolicited emails from legitimate ones. People had to sift through their emails to identify spam manually. Nowadays, email spam filters have automated this task and are quite efficient at identifying spam emails and keeping only ham (nonspam) emails in the box. The Bayesian approach forms the principle behind many spam filters. Consider the following example:

Question: What is the probability of a mail being spam, given that it contains the word "offer"? Available data indicates that 50% of all emails are spam mails. 9% of spam emails contain the word "offer," and 0.4 % of ham emails contain the word "offer."

Answer:

Defining the events and probabilities as follows:

A: Email is "spam"

~A: Email is "ham"

B: Email contains the word "offer"

P(A) = 0.5 (assuming 50% of emails are spam mails)

P(B/A) = Probability of spam mail containing the word "offer" = 0.09 (9%)

P(B/~A) = Probability of ham mail containing the word "offer" = 0.004 (0.4%)

Applying the Bayes theorem:

P(A/B) = (0.09*0.5)/(0.09*0.5)+(0.004)*(0.5) = 0.957

In other words, the probability of the mail being a spam mail given that it has the word "offer" is 0.957.

SciPy library

Scipy, a library for mathematical and scientific computations, contains several functions and algorithms for a wide range of domains, including image processing, signal processing, clustering, calculus, matrices, and statistics. Each of these areas has a separate submodule in SciPy. We use the *scipy.stats* submodule in this chapter, and apply the functions from this submodule for statistical tests and different types of distributions. This module also contains functions for distance calculations, correlations, and contingency tables.

Further reading:

Read more about the *scipy.stats* module and its functions:

`https://docs.scipy.org/doc/scipy/reference/stats.html`

Probability distributions

To understand probability distributions, let us first look at the concept of random variables, which are used to model probability distributions.

Random variable: A variable whose values equal the numeric values associated with the outcomes of a random experiment.

Random variables are of two types:

1. Discrete random variables can take a finite, countable number of values. For example, a random variable for the Likert scale, used for surveys and questionnaires to assess responses, can take values like 1, 2, 3, 4, and 5. The *probability mass function*, or *PMF*, associated with a discrete random variable is a function that provides the probability that this variable is exactly equal to a certain discrete value.

2. Continuous random variables can take infinitely many values. Examples include temperature, height, and weight. For a continuous variable, we cannot find the absolute probability. Hence, we use the *probability density function*, or *PDF*, for continuous variables (the equivalent of PMF for discrete variables). The PDF is the probability that the value of a continuous random variable falls within a range of values.

   The cumulative distribution function (CDF) gives the probability of a random variable being less than or equal to a given value. It is the integral of the PDF and gives the area under the curve defined by the PDF up to a certain point.

   In the following section, we cover the two types of probability distributions for discrete random variables: binomial and Poisson.

Binomial distribution

In a binomial experiment, there are several independent trials, with every trial having only *two possible outcomes*. These outcomes are the two values of the binomial discrete random variable. A basic example of a binomial distribution is the repeated toss of a coin. Each toss results in only two outcomes: Heads or Tails.

The following are the characteristics of a binomial distribution:

1. There are *n* identical trials
2. Each trial results in either one of only two possible outcomes
3. The outcomes of one trial do not affect the outcomes of other trials
4. The probability of success (*p*) and failure (*q*) is the same for each trial
5. The random variable represents the number of successes in these *n* trials and can at most be equal to *n*
6. The mean and variance of the binomial distribution are as follows:

   Mean = *n***p* (number of trials*probability of success)

   Variance = *n***p***q* (number of trials*probability of success*probability of failure)

The PMF, or the probability of *r* successes in *n* attempts of an experiment, is given by the following equation:

$$P(X=r)= {}^{n}C_{r}p^{r}q^{n-r}$$

Where *p* is the probability of success, *q* is the probability of failure, and *n* is the number of trials

The shape of a binomial distribution

The binomial distribution resembles a skewed distribution, but it approaches symmetry and looks like a normal curve as *n* increases and *p* becomes smaller, as shown in Figure 9-2.

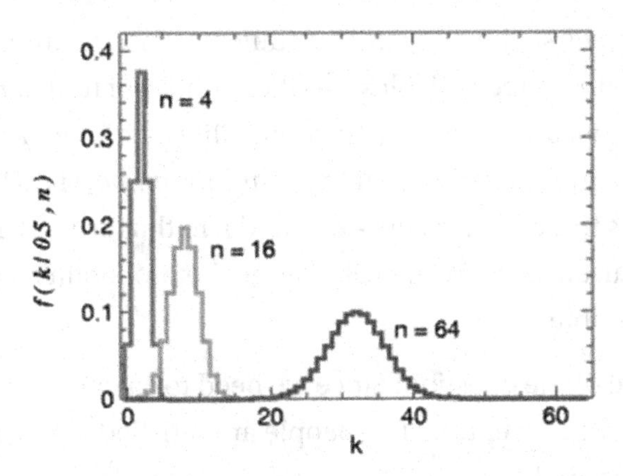

***Figure 9-2.** Binomial distribution for different values*

Question: The metro rail company surveys eight senior citizens traveling in a subway train about their satisfaction with the new safety features introduced in the subway trains. Each response has only two values: yes or no. Let us assume that the probability of a "yes" response is 0.6, and the probability of a "no" response is 0.4 based on historical survey.

Calculate the probability that

1. Exactly three people are satisfied with the metro's new safety features
2. Fewer than five people are satisfied

Solution:

1. For part 1 of the question: We can either use the formula ${}^{n}C_{r}p^{r}q^{n-r}$ or solve it using a Scipy function (*stats.binom.pmf*), as shown in the following:

   CODE:

   ```
   import scipy.stats as stats
   n,r,p=8,3,0.6
   stats.binom.pmf(r,n,p)
   ```

   Output:

   ```
   0.12386304000000009
   ```

Explanation: First, the *scipy.stats* module needs to be imported. Then we define three variables – *n* (the number of trials), *r* (the number of successes), and *p* (the probability of failure). After this, the PMF for binomial distributions (*stats.binom.pdf*) is called, and we pass three parameters - *r, n,* and *p* in that order. The *pmf* function is used since we are calculating the probability of a discrete variable.

2. For part two of the question: Since we need to calculate the probability that fewer than five people are satisfied, the limiting value of the variable is 4.

   The following equation gives the probability we need to calculate:

$$P(X<=4)=P(X=0)+P(X=1)+P(X=2)+P(X=3)+P(X=4)$$

   We can either apply the formula $^nC_rp^rq^{n-r}$ to calculate the values for $r = 0, 1, 2, 3,$ and 4 or solve using the *stats.binom.cdf* function in Scipy as follows:

   CODE:

```
import scipy.stats as stats
n,r,p=8,4,0.6
stats.binom.cdf(r,n,p)
```

   Output:

```
0.4059135999999976
```

   Explanation: We use the CDF function when we calculate the probability for more than one value of x.

Poisson distribution

A Poisson distribution is a distribution that models the number of events that occur over a given interval (usually of time, but can also be an interval of distance, area, or volume). The average rate of occurrence of events needs to be known.

The PMF for a Poisson distribution is given by the following equation:

$$P(x=r)=\frac{\lambda^r e^{-\lambda}}{r!}$$

where P(x=r) is the probability of the event occurring r number of times, r is the number of occurrences of the event, and λ^r represents the average/expected number of occurrences of that event.

The Poisson distribution can be used to calculate the number of occurrences that occur over a given period, for instance:

- number of arrivals at a restaurant per hour
- number of work-related accidents occurring at a factory over a year
- number of customer complaints at a call center in a week

Properties of a Poisson distribution:

1. Mean=variance=λ. In a Poisson distribution, the mean and variance have the same numeric values.
2. The events are independent, random, and cannot occur at the same time.
3. When *n* is >20 and *p* is <0.1, a Poisson distribution can approximate the binomial distribution. Here, we substitute $\lambda = np$.
4. When the value of *n* is large, *p* is around 0.5, and *np* > 0.5, a normal distribution can be used to approximate a binomial distribution.

The shape of a Poisson distribution

A Poisson distribution is skewed in shape but starts resembling a normal distribution as the mean (λ) increases, as shown in Figure 9-3.

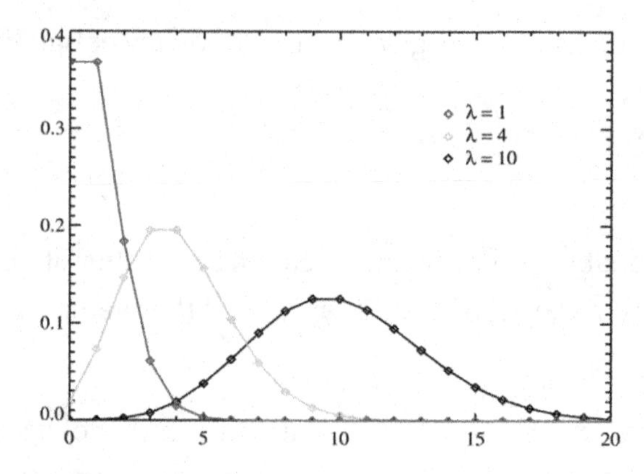

***Figure 9-3.** Poisson distribution*

Solved example for the Poisson distribution:

In a subway station, the average number of ticket-vending machines out of operation is two. Assuming that the number of machines out of operation follows a Poisson distribution, calculate the probability that a given point in time:

1. Exactly three machines are out of operation
2. More than two machines are out of operation

Solution:

1. We can either use the formula $\frac{\lambda^r e^{-\lambda}}{r!}$ or solve it in Python as follows:

   CODE:

   ```
   import scipy.stats as stats
   λ=2
   r=3
   stats.poisson.pmf(r,λ)
   ```

   Output:

   ```
   0.18044704431548356
   ```

   Explanation: First, the *scipy.stats* module needs to be imported. Then we define two variables - λ (the average) and *r* (the number of occurrences of the event). Then, the PMF for a Poisson distribution (*stats.poisson.pmf*) is called, and we pass the two arguments to this function, *r* and λ, in that order.

2. Since we need to calculate the probability that more than two machines are out of order, we need to calculate the following probability:

   P(x>2), or (1-p(x=0)-p(x=1)-p(x=2)).

   This can be computed using the *stats.poisson.cdf* function, with r=2.

   CODE:

```
import scipy.stats as stats
λ=2
r=2
1-stats.poisson.cdf(r,λ)
```

   Output:

```
0.3233235838169366
```

Explanation: We follow a similar method as we did for the first part of the question but use the CDF function (*stats.poisson.cdf*) instead of PMF (*stats.poisson.pmf*).

Continuous probability distributions

There are several continuous probability distributions, including the normal distribution, Student's T distribution, the chi-square, and ANOVA distribution. In the following section, we explore the normal distribution.

Normal distribution

A normal distribution is a symmetrical bell-shaped curve, defined by its mean (μ) and standard deviation (σ), as shown in Figure 9-4.

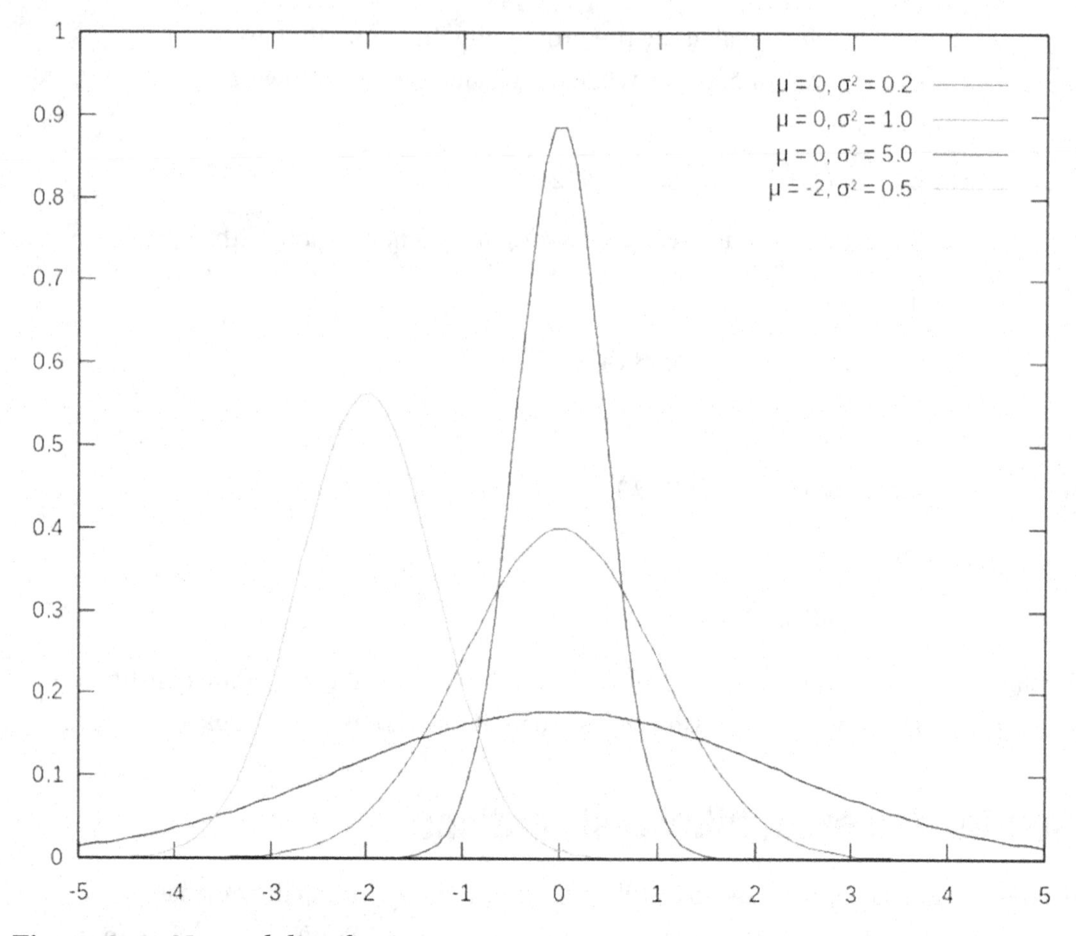

***Figure 9-4.** Normal distribution*

All the four curves in Figure 9-4 are normal distributions. The mean is represented by the symbol μ (mu) and the standard deviation by the symbol σ (sigma)

Characteristics of a normal distribution

1. The central value (μ) is also the mode and the median for a normal distribution

2. Checking for normality: In a normal distribution, the difference between the 75th percentile value (Q3) and the 50th percentile value (median or Q2) equals the difference between the median (Q2) and the 25th percentile. In other words,

$$Q_3 - Q_2 = Q_2 - Q_1$$

If the distribution is skewed, this equation does not hold.

In a right-skewed distribution, $(Q_3 - Q_2) > (Q_2 - Q_1)$

In a left-skewed distribution, $(Q_2 - Q_1) > (Q_3 - Q_2)$

Standard normal distribution

To standardize units and compare distributions with different means and variances, we use a standard normal distribution.

Properties of a standard normal distribution:

- The standard normal distribution is a normal distribution with a mean value of 0 and a standard deviation as 1.
- Any normal distribution can be converted into standard normal distribution using the following formula:
  $z = \frac{(x - \mu)}{\sigma}$, where μ and σ are the mean and variance of the original normal distribution.
- In a standard normal distribution,
  - 68.2% of the values lie within 1 standard deviation of the mean
  - 95.4% of the values lie between 2 standard deviations of the mean
  - 99.8% lie within 3 standard deviations of the mean

This distribution of values is shown in Figure 9-5.

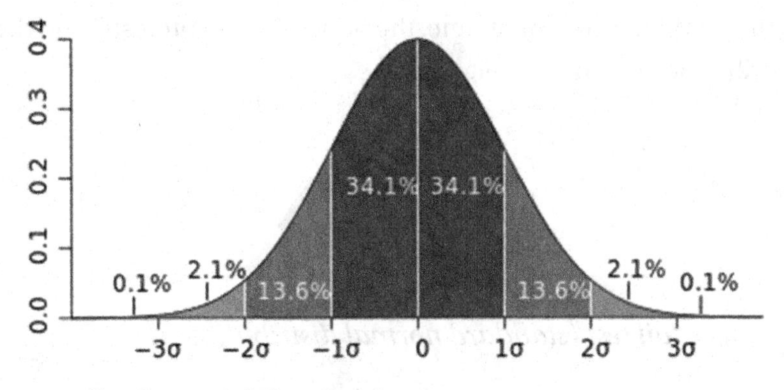

Figure 9-5. *Standard normal distribution*

- The area under the standard normal distribution between any two points represents the proportion of values that lies between these two points. For instance, the area under the curve on either side of the mean is 0.5. Put in another way, 50% of the values lie on either side of the mean.

There are two types of questions involving normal distributions:

1. Calculating probability/proportion corresponding to the value of a variable: The z-value is calculated using the formula $z=\frac{(x-\mu)}{\sigma}$, and this z-value is then passed as an argument to the *stats.norm.cdf* function
2. Calculating the value of the variable corresponding to a given probability: First, the z-value is obtained by passing the probability value as an argument to the *stats.norm.ppf* function. Then, we obtain the value of the variable (x) corresponding to the z-value by substituting values in the following formula: $z=\frac{(x-\mu)}{\sigma}$

Solved examples: Standard normal distribution

Question: An IT team in a software company is inspecting some laptops. The team needs to select the top 1% of the laptops, with the criterion being the fastest boot times. The average boot time is 7 seconds, with a standard deviation of 0.5 seconds. What would be the cutoff boot time necessary for selection?

Solution:

Step 1: Since the criterion is fast boot time, the boot times of interest lie on the lower left end of the distribution, as shown in Figure 9-6.

Figure 9-6. *Lower-tail test (standard normal distribution)*

The area of the curve to the right of this tail is 0.99. We calculate the z-value, corresponding to a probability value of 0.99, using the *stats.norm.ppf* function:

CODE:

```
stats.norm.ppf(0.99)
```

Output:

```
2.3263478740408408
```

Since this is a lower-tail test, we take the value of z as –2.33 (this value is negative as it lies to the left of the mean). We can also verify this using the z-table by calculating the z-value corresponding to a probability of 0.99.

Step 2: Apply the following formula and calculate x

$$z=(x-\mu)/\sigma$$

where $z = -2.33$, $\mu = 7$, and $\sigma = 0.5$. We need to calculate the value of x:

CODE:

```
x=(-2.33*0.5)+7
```

Output: 5.835

Inference: The required boot time is 5.835 seconds

Example 2 (standard normal distribution):

A company manufactures tube lights, where the life (in hours) of these tube lights follows a normal distribution with a mean of 900 (hrs) and a standard deviation of 150 (hrs). Calculate the following:

(1) The proportion of tube lights that fail within the first 750 hours

(2) The proportion of tube lights that fail between 800 and 1100 hours

(3) After how many hours would 20% of the tube lights fail?

Solution for Example 2 (standard normal distribution):

(1) Calculate the z-value corresponding to X=750 and obtain the corresponding probability:

CODE:

```
x=750
μ=900
σ=150
z=(x-μ)/σ
z
stats.norm.cdf(z)
```

Output:

```
0.15865525393145707
```

Inference: 15.8% of the tube lights fail within the first 750 hours.

(2) Calculate the z-values corresponding to x-values of 800 and 1100, respectively, and subtract the probabilities corresponding to these z-values.

CODE:

```
x1=800
x2=1100
μ=900
σ=150
z1=(x1-μ)/σ
z2=(x2-μ)/σ
p2=stats.norm.cdf(z2)
p1=stats.norm.cdf(z1)
p2-p1
```

Output:

```
0.6562962427272092
```

Inference: Around 65.6% of the tube lights, with a lifetime between 800 and 1100 hours, fail.

(3) Calculate the z-value corresponding to a probability of 0.2 and calculate x by substituting z, μ, and σ in the formula z= $\frac{(x-\mu)}{\sigma}$

CODE:

```
z=stats.norm.ppf(0.2)
μ=900
σ=150
x=μ+σ*z
x
```

Output:

```
773.7568149640629
```

Inference: After a lifetime of around 774 hours, 20% of the tube lights fail.

Measures of central tendency

The central tendency is a measure of the central value among a set of values in a dataset. The following are some of the measures of central tendency:

Mean: This is the average of values in a dataset.

Median: This is the middle number when the values in the dataset are arranged size-wise.

Mode: The most frequently occurring value in a dataset with discrete values.

Percentile: A percentile is a measure of the percentage of values below a particular value. The median corresponds to the 50th percentile.

Quartile: A quartile is a value that divides the values in an ordered dataset into four equal groups. Q1 (or the first quartile) corresponds to the 25th percentile, Q2 corresponds to the median, and Q3 corresponds to the 75th percentile.

Measures of dispersion

The measures of dispersion give a quantitative measure of the spread of a distribution. They provide an idea of whether the values in a distribution are situated around the central value or spread out. The following are the commonly used measures of dispersion.

Range: The range is a measure of the difference between the lowest and highest values in a dataset.

Interquartile range: A measure of the difference between the third quartile and the first quartile. This measure is less affected by extreme values since it focuses on the values lying in the middle. The interquartile range is a good measure for skewed distributions that have outliers. The interquartile range is denoted by IQR = Q3 - Q1.

Variance: This is a measure of how much values in a dataset are scattered around the mean value. The value of the variance is a good indication of whether the mean is representative of values in the dataset. A small variance would indicate that the mean is an appropriate measure of central tendency. The following formula gives the variance:

$$\sigma^2 = \frac{\Sigma(x-\mu)^2}{N},$$

Where μ is the mean, and N is the number of values in the dataset.

Standard deviation: This measure is calculated by taking the square root of the variance. The variance is not in the same units as the data since it takes the square of the differences; hence taking the square root of the variance brings it to the same units as the data. For instance, in a dataset about the average rainfall in centimeters, the variance would give the value in cm^2, which would not be interpretable, while the standard deviation in cm would give an idea of the average rainfall deviation in centimeters.

Measures of shape

Skewness: This measures the degree of asymmetry of a distribution, as shown in Figure 9-7.

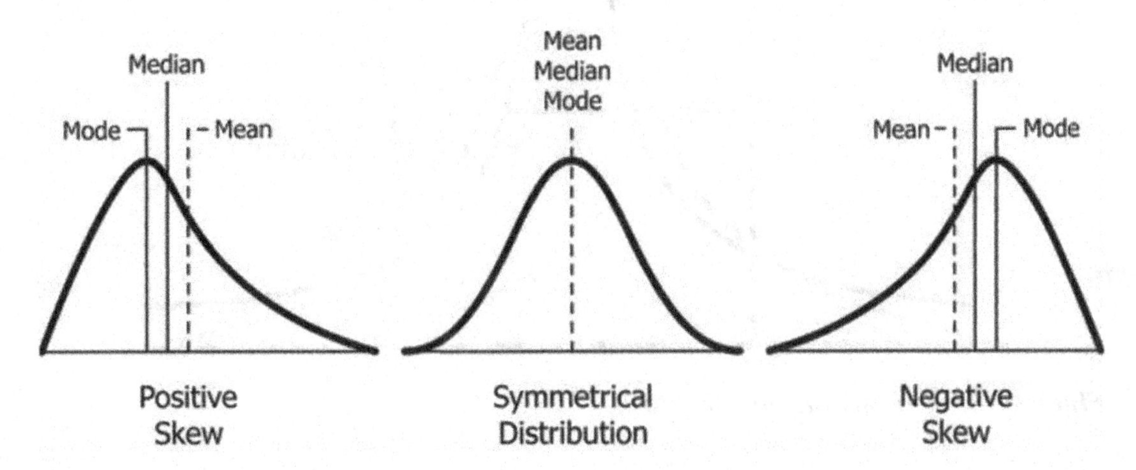

Figure 9-7. *Distributions with varied skewness*

We can observe the following from the Figure 9-7:

In a positively skewed distribution, mean > median

In a negatively skewed distribution, mean < median

In a perfectly symmetrical distribution, mean = median = mode

Kurtosis

Kurtosis is a measure of whether a given distribution of data is curved, peaked, or flat.

A mesokurtic distribution has a bell-shaped curve. A leptokurtic distribution is one with a marked peak. A platykurtic distribution, as the name indicates, has a flat curve. These distributions are shown in Figure 9-8.

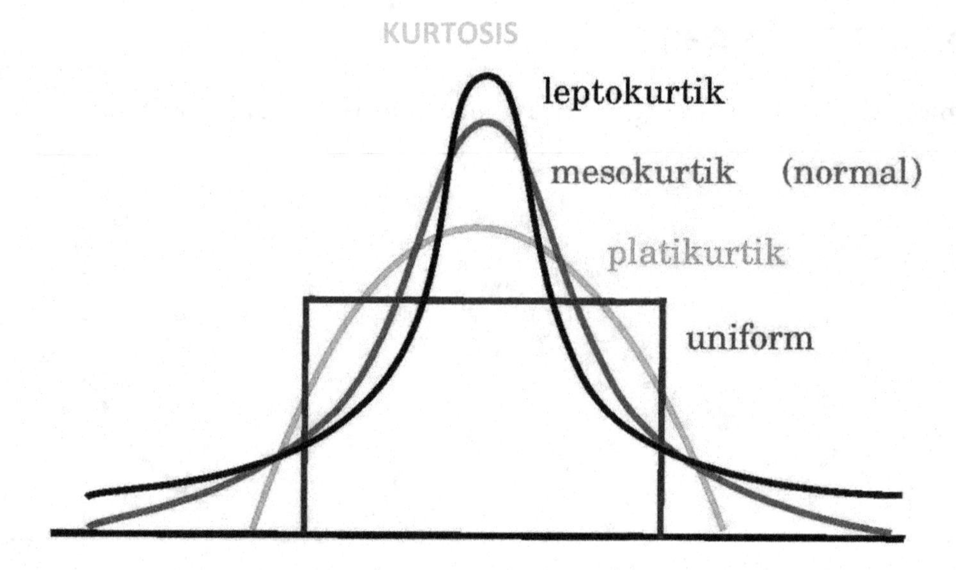

Figure 9-8. *Representation of kurtosis*

Solved Example:

The weight of children (in kgs) aged 3-7 in a primary school is as follows: 19, 23, 19, 18, 25, 16, 17, 19, 15, 23, 21, 23, 21, 11, 6. Let us calculate the measures of central tendency, dispersion, skewness, and kurtosis.

Creating a Pandas Series object:

CODE:

```
import pandas as pd
a=pd.Series([19,23,19,18,25,16,17,19,15,23,21,23,21,11,6])
```

The Pandas *describe* method can be used with either the series object or the DataFrame object and is a convenient way of obtaining most of the measures of central tendency in one line of code. The mean is 18.4 kgs, the first quartile (Q1 or 25th percentile) is 16.5 kgs, the median (50th percentile) is 19 kgs, and the third quartile (75th percentile) is 22 kgs.

CODE:

```
a.describe()
```

Output:

```
count    15.000000
mean     18.400000
std       4.997142
min       6.000000
25%      16.500000
50%      19.000000
75%      22.000000
max      25.000000
dtype: float64
```

Obtain the mode using the *mode* method:

CODE:

```
a.mode()
```

Output:

```
0    19
1    23
dtype: int64
```

The values 19 and 23 are the most frequently occurring values.

Obtain the measures of dispersion

The range can be calculated using the *max* and *min* functions and taking the difference between these two values:

CODE:

```
range_a=max(a)-min(a)
range_a
```

Output:

```
19
```

Obtain the standard deviation and variance using the *std* and *var* methods, respectively:

CODE:

```
a.std()
```

Output:

```
4.99714204034952
```

CODE:

```
a.var()
```

Output:

```
24.97142857142857
```

Obtain the measures of skewness and kurtosis by using the *skew* and *kurtosis* functions from the *scipy.stats* module:

CODE:

```
from scipy.stats import skew,kurtosis
stats.kurtosis(a)
```

Output:

```
0.6995494033062934
```

A positive value of kurtosis means that the distribution is leptokurtic.

Skewness:

CODE:

```
stats.skew(a)
```

Output:

```
-1.038344732097918
```

A negative value of skewness implies that the distribution is negatively skewed, with the mean less than the median.

Points to note:

1. The mean value is affected by outliers (extreme values). Whenever there are outliers in a dataset, it is better to use the median.
2. The standard deviation and variance are closely tied to the mean. Thus, if there are outliers, standard deviation and variance may not be representative measures too.
3. The mode is generally used for discrete data since there can be more than one modal value for continuous data.

Sampling

When we try to find out something about a population, it is not practical to collect data from every subject in the population. It is more feasible to take a sample and make an estimate of the population parameters based on the sample data. The sample's characteristics should align with that of the population. The following are the main methods of collecting samples.

Probability sampling

Probability sampling involves a random selection of subjects from the population. There are four main methods of doing probability sampling:

1. Simple random sampling: Subjects are chosen randomly, without any preference. Every subject in the population has an equal likelihood of being selected.
2. Stratified random sampling: The population is divided into mutually exclusive (non-overlapping) groups, and then subjects are randomly selected from each group. Example: If you are surveying to assess preference for subjects in a school, you may divide students into male and female groups and randomly select subjects from each group. The advantage of this method is that it represents all categories or groups in the population.

3. Systematic random sampling: Subjects are chosen at regular intervals. Example: To take a sample of 100 people from a population of 500, first divide 500 by 100, which equals 5. Now, take every 5th person for our sample. It is easier to perform but may not be representative of all the subjects in the population.

4. Cluster sampling: Here, the population is divided into non-overlapping clusters covering the entire population between them. From these clusters, a few are randomly selected. Either all the members of the chosen clusters are selected (one-stage), or a subset of members from the selected clusters is randomly chosen (two-stage). The advantage of this method is that it is cheaper and more convenient to carry out.

Non-probability sampling

When it is not possible to collect samples using probability sampling methods due to a lack of readily available data, we use non-probability sampling techniques. In non-probability sampling, we do not know the probability of a subject being chosen for the study.

It is divided into the following types:

1. Convenience sampling: Subjects that are easily accessible or available are chosen in this method. For example, a researcher can select subjects for their study from their workplace or the university where they work. This method is easy to implement but may not be representative of the population.

2. Purposive: Subjects are chosen based on the purpose of the sampling. For example, if a survey is being carried out to assess the effectiveness of intermittent fasting, it needs to consider the age group of the population that can undergo this fast, and the survey may only include people aged 25-50. Purposive sampling is further divided into

    - Quota sampling: Quotas are taken in such a way that the significant characteristics of the population are taken into account while samples are chosen. If a population has 60% Caucasians, 20% Hispanics, and 20% Asians, the sample you choose should have the same percentages.

- Snowball sampling: In this method, the researcher identifies someone they know who meets the criteria of the study. This person then introduces to others the person may know, and the sample group thus grows through word-of-mouth. This technique may be used for populations that lack visibility, for example, a survey of people suffering from an under-reported illness.

Central limit theorem

The central limit theorem states that if we choose samples from a population, the means of the samples are normally distributed, with a mean as μ and standard deviation as $\sigma_{\bar{x}}$.

Even if the population distribution is not a normal distribution by itself, the distribution of the sample means resembles a normal distribution. As the sample size increases, the distribution of sample means becomes a closer approximation to the normal distribution, as seen in Figure 9-9.

Histograms of 500 Observed Sample Means Randomly Drawn from a Population (0 to 100) with a Uniform Distribution for Various Sample Sizes (*N*)

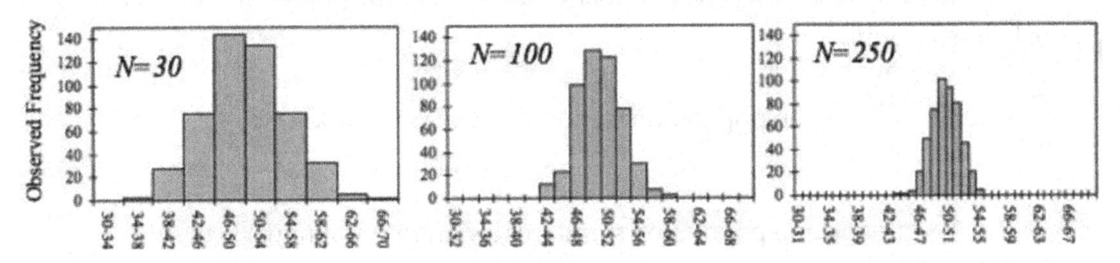

***Figure 9-9.** Distribution of sample means. As the sample size increases, the distribution of sample means resembles a normal distribution*

The sample mean is used as an estimate for the population mean, but the standard deviation of this sampling distribution ($\sigma_{\bar{x}}$), is not the same as the population standard deviation, σ. The sample standard deviation is related to the population standard deviation as follows:

$$\sigma_{\bar{x}} = \frac{\sigma}{\sqrt{n}}$$

where σ is the population standard deviation and n is the sample size

$\sigma_{\bar{x}}$ is known as the standard error (of the distribution of sample means). As the sample size (n increases), the standard error approaches 0, and the sample mean ($\bar{x}$) approaches the population mean (μ).

Estimates and confidence intervals

Point estimate: A single statistic extracted from a sample that is used to estimate an unknown population parameter. The sample mean is used as a point estimate for the population mean.

Interval estimate: The broad range of values within which the population parameter lies. It is indicative of the error in estimating the population parameter.

Confidence interval: The interval within which the value of the population mean lies. For a random sample of size n and mean $\bar{x}$ taken from a population (with standard deviation as σ, and mean as μ), the confidence interval for the population mean is given by the following equations:

$\bar{x} - \frac{z\sigma}{\sqrt{n}} \leq \mu \leq \bar{x} + \frac{z\sigma}{\sqrt{n}}$: when population standard deviation, σ, is known

$\bar{x} - \frac{zs}{\sqrt{n}} \leq \mu \leq \bar{x} + \frac{zs}{\sqrt{n}}$: when population standard deviation is unknown (s in this equation is the sample standard deviation)

Solved example: Confidence intervals

Question: A sample (consisting of ten subjects) is taken from a certain population of students. The grade point averages of these students are normally distributed. The population standard deviation is not known. Calculate the 95% confidence interval for the population mean (grade point average for the whole student population), based on the following sample values: 3.1, 2.9, 3.2, 3.4, 3.7, 3.9, 3.9, 2.8, 3.4, 3.6.

Solution:

The following code calculates the 95% confidence interval for the population mean:

CODE:

```
import numpy as np
import scipy.stats as stats
grades = np.array([3.1,2.9,3.2,3.4,3.7,3.9,3.9,2.8,3.4,3.6])
stats.t.interval(0.95, len(grades)-1, loc=np.mean(grades), scale=stats.
sem(grades))
```

Output:

```
(3.1110006165952773, 3.668999383404722)
```

Interpretation: There is a 95% probability that the grade point average for the population of students falls between 3.11 and 3.67.

Explanation of the preceding code: We first define a NumPy array for the sample observations, and then call the *stats.t.interval* function. To this function, we pass the following arguments: the values of the confidence interval (0.05), degrees of freedom (total number of observations: 1), the sample mean, and standard error (calculated by the function *stats.sem*). The function returns two values – the lower confidence interval (LCI) and the upper confidence interval (UCI). Note that the *stats.t.interval* function is used because the population standard deviation is not known. If it were known, we would use the function *stats.norm.interval.*

Types of errors in sampling

If we take a sample and make inferences about the entire population based on this sample, errors inevitably arise. These errors can broadly be classified as follows:

- Sampling error: Difference between the sample estimate for the population and the actual population estimate
- Coverage error: Occurs when the population is not adequately represented, and some groups are excluded

- Nonresponse errors: Occurs when we fail to include nonresponsive subjects that satisfy the criteria of the study, but are excluded since they do not answer the survey questions
- Measurement error: Not measuring the correct parameters due to flaws in the method or tool used for measurement

We now move on to concepts in hypothesis testing.

Hypothesis testing

A hypothesis is a statement that gives the estimate of an unknown variable or parameter. If we are trying to find the average age of people in a city from a sample drawn from this population, and we find that the average age of people in this sample is 34, our hypothesis statement could be as follows: "The average age of people in this city is 34 years."

Basic concepts in hypothesis testing

In a hypothesis test, we construct two statements known as the null and alternate hypothesis.

Null hypothesis: Denoted by the term H_0,this is the hypothesis that needs to be tested. It is based on the principle that there is no change from the status quo. If the sample mean is 70, while the historical population mean is 90, the null hypothesis would state that the population mean equals 90.

Alternate hypothesis: Denoted by the term H_1, this hypothesis is what one would believe if the null hypothesis does not hold. The alternate hypothesis (using the preceding example) would state that the mean is greater than, less than, or not equal to 90.

We either reject the null hypothesis or fail to reject the null hypothesis. Note that rejecting the null hypothesis does not imply that the alternative hypothesis is true. The result of a hypothesis test is only suggestive or indicative of something regarding the population, and it does not conclusively prove or disprove any hypothesis.

Key terminology used in hypothesis testing

Let us look at some commonly used terms in hypothesis testing:

Type 1 error, or the **level of significance**, denoted by the symbol α, is the error of rejecting the null hypothesis when it is true. It can also be defined as the probability that the population parameter lies outside its confidence interval. If the confidence interval is 95%, the level of significance is 0.05, or there is a 5% chance that the population parameter does not lie within the confidence interval calculated from the sample.

Example of a Type 1 error: Mr. X has a rash and goes to a doctor to get a test for chickenpox. Let the null hypothesis be that he does not have this illness. The doctor incorrectly makes a diagnosis for chickenpox based on some faulty tests, but the reality is that Mr. X does not have this illness. This is a typical example of rejecting the null hypothesis when it is true, which is what a Type 1 error is.

Type 2 error, denoted by the symbol β,is the error that occurs when the null hypothesis is not rejected when it is false. In the preceding chickenpox example, if Mr. X suffers from chickenpox, but the doctor does not diagnose it, the doctor is making a Type 2 error.

One-sample test: This is a test used when there is only one population under consideration, and a single sample is taken to see if there is a difference between the values calculated from the sample and population parameter.

Two-sample test: This is a test used when samples are taken from two different populations. It helps to assess whether the population parameters are different based on the sample parameters.

The critical test statistic: The limiting value of the sample test statistic to decide whether or not to reject the null hypothesis. In Figure 9-10, z=1.96 and z=-1.96 are critical values. Z-values greater than 1.96 and less than -1.96 lead to rejection of the null hypothesis.

Region of rejection: The range of values where the null hypothesis is rejected. The region of acceptance is the area corresponding to the limits where the null hypothesis holds.

The regions of rejection and acceptance are shown in Figure 9-10.

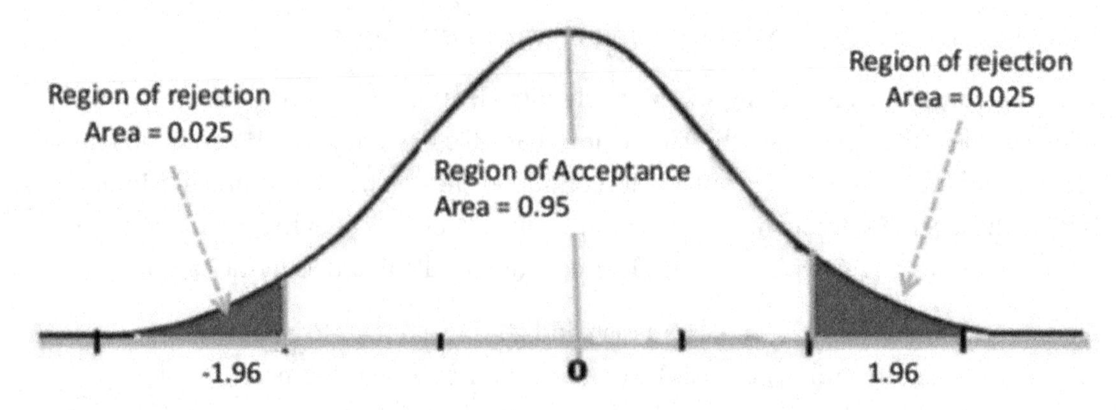

***Figure 9-10.** Regions of acceptance and rejection*

Two-tail test: The region of rejection is located on both tails of the distribution.

Example of a two-tail test: A sample of 10 students is taken from a class of 50 students to see if there is a change in the mean score of the class with respect to its historical average. This is an example of a case where we will conduct a two-tail test because we are just testing for a change in the mean, and we do not know if this change is positive or negative.

One-tail test: The region of rejection is located on the right tail (upper one-tail) or the left tail (lower-tail) but not on both tails.

Example of upper-tail test: Afterschool classes are being conducted to improve scores in a class of 50 students. These special classes are believed to have improved the scores. To test this hypothesis, we perform a one-tail test (upper) using a sample from the population because we are testing if the mean score has increased. The region of rejection will be located on the right tail.

Example of lower-tail test: Due to political unrest, there has been an increase in absenteeism among students. It is believed that these events may negatively affect the scores of the students. To test this hypothesis, we conduct a one-tail test (lower) using a sample from the population because we are testing if the mean score has reduced. The region of rejection is located on the left tail.

The p-value (denoted by the letter p) is the probability of obtaining a value of the test statistic at least as extreme as the one observed, assuming the null hypothesis is true.

The p-value is often used in hypothesis tests to decide whether or not to reject a null hypothesis. The p-value is commonly compared with a significance level of 0.05.

If $p < 0.05$, it would mean that the probability that the sample data was random and not representative of the population is very low. We reject the null hypothesis in this case.

If $p > 0.05$, there is a greater chance that this sample is not representative of the population. We fail to reject the null hypothesis in this case.

Steps involved in hypothesis testing

1. State the null and alternate hypothesis
2. Fix the level of significance and obtain the critical value of the test statistic
3. Select the appropriate test:

   Choose the test based on the following parameters:

   - Number of samples
   - Whether the population is normally distributed
   - The statistic being tested
   - The sample size
   - Whether the population standard deviation is known
4. Obtain the relevant test statistic (z statistic/t statistic/chi-square statistic/f statistic) or the p-value
5. Compare the critical test statistic with the calculated test static or compare the p-value with 0.05

   Reject the null hypothesis based on either the test statistic or the p-value:

   - Using the test statistic:
     - calculated test static>critical test statistic (upper-tail test)
     - calculated test static<critical test statistic (lower-tail test)

OR

- Using the p-value (p) if $p<0.05$

6. Draw an inference based on the preceding comparison

One-sample z-test

This test is used when we want to verify if the population mean differs from its historical or hypothesized value.

Criteria for a one-sample z-test:

- The population from which the sample is drawn is normally distributed
- The sample size is greater than 30
- A single sample is drawn
- We are testing for the population mean
- The population standard deviation is known

Formula for calculating test statistic: $z = \frac{(\bar{x} - \mu)}{\sigma / \sqrt{n}}$,

where $\bar{x}$ is the sample mean, μ is the population mean, σ is the population standard deviation, and n is the sample size

Solved example: One-sample z-test

Question: A local Italian restaurant has an average delivery time of 45 minutes with a standard deviation of 5 minutes. The restaurant has received some complaints from its customers and has decided to analyze the last 40 orders. The average delivery time for these 40 orders was found to be 48 minutes. Conduct the appropriate test at a significance level of 5% to decide whether the delivery times have increased.

Answer:

1. State the hypothesis:

   Let μ be the average delivery time for the restaurant (population mean)
   Null hypothesis: H_0: μ=45
   Alternate hypothesis: $H_1 : \mu > 45$

2. Fix the level of significance: α=0.05

3. Select the appropriate hypothesis test:

   - Number of samples: 1
   - Sample size: n=40 (Large)
   - What we are testing: Whether there is a difference between the sample mean ($\overline{x} = 48$) and the population mean (μ=45)
   - Population standard deviation (σ=5) is known

   We select the one-sample z-test based on the preceding data.

4. Obtain the test statistic and p-value, with the help of the following equation:

$$z = \frac{(\overline{x} - \mu)}{\sigma / \sqrt{n}}$$

   Substituting the values $\overline{x} = 48$, μ=45, $\sigma = 5$, and n=40:

   z=3.7947

   Calculate the p-value corresponding to this z-value using the *stats.norm.cdf* function:

   CODE:

```
import scipy.stats as stats
stats.norm.cdf(z)
```

Output:

```
0.999
```

5. Compare the p-value with the level of significance (0.05):

   Since the calculated p-value is $>\alpha$, we fail to reject the null hypothesis.

6. Inference:

   There is no significant difference, at a level of 0.05, between the average delivery time of the sample and the historical population average.

Two-sample sample z-test

A two-sample z-test is similar to a one-sample z-test, the only differences being as follows:

- There are two groups/populations under consideration and we draw one sample from each population
- Both the population distributions are normal
- Both population standard deviations are known
- The formula for calculating test statistic: : $z = \dfrac{(\bar{x}_1 - \bar{x}_2)}{\sqrt{\left[\dfrac{\sigma_1^2}{n_1} + \dfrac{\sigma_2^2}{n_2}\right]}}$

Solved example: Two-sample sample z-test

An organization manufactures LED bulbs in two production units, A and B. The quality control team believes that the quality of production at unit A is better than that of B. Quality is measured by how long a bulb works. The team takes samples from both units to test this. The mean life of LED bulbs at units A and B are 1001.3 and 810.47, respectively. The sample sizes are 40 and 44. The population variances are known: σ_A^2=48127 and σ_B^2=59173.

Conduct the appropriate test, at 5% significance level, to verify the claim of the quality control team.

Solution:

1. State the hypothesis:

   Let the mean life of LED bulbs at unit A and B be μ_A *and* μ_B, respectively.
   Null hypothesis: H_0: $\mu_A \leq \mu_B$
   Alternate hypothesis: $H_1 : \mu_A > \mu_B$
   This is a one-tail (upper-tail) test

2. Fix the level of significance: α=0.05

3. Select the appropriate hypothesis test:

   - Number of samples: 2 samples (taking samples from two different populations)
   - Sample size: Large ($n_A = 40$, and $n_B = 44$)
   - What we are testing: Comparing the mean lifetime of LED bulbs in unit A with that of unit B
   - Population characteristics: The distribution of population is not known, but population variances are known
   - Hence, we conduct the two-sample z-test.

4. Calculate the test statistic and p-value

   Use the following equation:

$$z = \frac{(\bar{x}_1 - \bar{x}_2)}{\sqrt{\left[\frac{\sigma_1^2}{n_1} + \frac{\sigma_2^2}{n_2}\right]}}$$

   Substituting the values $\bar{x}_1 = 1001.3, \bar{x}_2 = 810.47$, $n_1 = 40$, $n_2 = 44$ and the variance(sigma) values of 48127 and 59173 in the preceding formula to calculate z:

   CODE:

```
z=(1001.34-810.47)/(48127/40+59173/44)**0.5
```

   Output:

```
3.781260568723408
```

Calculate the p-value corresponding to this z-value using the *stats.norm.cdf* function:

CODE:

```
import scipy.stats as stats
p=1-stats.norm.cdf(z)
p
```

Output:

```
7.801812433294586e-05
```

Explanation: Since this is an upper-tail test, we need to calculate the area/proportion of values in the right tail. Hence, we subtract the area calculated (*stats.norm.cdf*) from 1.

5. Comparing the calculated p-value with the level of significance:

   Since the calculated p-value $(0.000078) < \alpha(0.05)$, we reject the null hypothesis.

6. Inference: The LED bulbs produced at unit A have a significantly longer life than those at unit B, at a 5% level.

Hypothesis tests with proportions

Proportion tests are used with nominal data and are useful for comparing percentages or proportions. For example, a survey collecting responses from a department in an organization might claim that 85% of people in the organization are satisfied with its policies. Historically the satisfaction rate has been 82%. Here, we are comparing a percentage or a proportion taken from the sample with a percentage/proportion from the population. The following are some of the characteristics of the sampling distribution of proportions:

- The sampling distribution of the proportions taken from the sample is approximately normal
- The mean of this sampling distribution ($\bar{p}$) = Population proportion (p)
- Calculating the test statistic: The following equation gives the z-value

$$z = \frac{(\bar{p} - p)}{\sqrt{\frac{p(1-p)}{n}}}$$

Where $\bar{p}$ is the sample proportion, p is the population proportion, and n is the sample size.

Solved example: One-sample proportion z-test

Here, we understand the one-sample proportion z-test using a solved example.

Question: It is known that 40% of the total customers are satisfied with the services provided by a mobile service center. The customer service department of this center decides to conduct a survey for assessing the current customer satisfaction rate. It surveys 100 of its customers and finds that only 30 out of the 100 customers are satisfied with its services. Conduct a hypothesis test at a 5% significance level to determine if the percentage of satisfied customers has reduced from the initial satisfaction level (40%).

Solution:

1. State the null and alternate hypothesis

   Let the average customer satisfaction rate be p

   H_0: $p = 0 \cdot 4$
   H_1: $p < 0 \cdot 4$

   The < sign indicates that this is a one-tail test (lower-tail)

2. Fix the level of significance: α=0.05

3. Select the appropriate test:

   We choose the one-sample z-test for proportions since

   - The sample size is large (100)
   - A single sample is taken
   - We are testing for a change in the population proportion

4. Obtain the relevant test statistic and p-value

$$z = \frac{(\bar{p} - p)}{\sqrt{\frac{p(1-p)}{n}}}$$

   Where $\bar{p} = 0.3, p = 0.4, n = 100$

Calculate z and p:

CODE:

```
import scipy.stats as stats
z=(0.3-0.4)/((0.4)*(1-0.4)/100)**0.5
p=stats.norm.cdf(z)
p
```

Output:

```
0.020613416668581790
```

5. Decide whether or not to reject the null hypothesis

   p-value (0.02)<0.05 → We reject the null hypothesis

6. Inference: At a 5% significance level, the percentage of customers satisfied with the service center's services has reduced

Two-sample z-test for the population proportions

Here, we compare proportions taken from two independent samples belonging to two different populations. The following equation gives the formula for the critical test statistic:

$$z = \frac{(\overline{p_1} - \overline{p_2})}{\sqrt{\frac{p_c(1-p_c)}{N_1} + \frac{p_c(1-p_c)}{N_2}}}$$

In the preceding formula, $\overline{p_1}$ is the proportion from the first sample, and $\overline{p_2}$ is the proportion from the second sample. $N_1$is the sample size of the first sample, and N_2 is the sample size of the second sample.

p_c is the pooled variance.

$$\overline{p_1} = \frac{x_1}{N_1}; \overline{p_2} = \frac{x_2}{N_2}; p_c = \frac{x_1 + x_2}{N_1 + N_2}$$

In the preceding formula, x_1 is the number of successes in the first sample, and x_2 is the number of successes in the second sample.

Let us understand the two-sample proportion test with the help of an example.

Question: A ride-sharing company is investigating complaints by its drivers that some of the passengers (traveling with children) do not conform with child safety guidelines (for example, not bringing a child seat or not using the seat belt). The company undertakes surveys in two major cities. The surveys are collected independently, with one sample being taken from each city. From the data collected, it seems that the passengers in City B are more noncompliant than those in City A. The law enforcement authority wants to know if the proportion of passengers conforming with child safety guidelines is different for the two cities. The data for the two cities is given in the following table:

	City A	City B
Total surveyed	200	230
Number of people compliant	110	106

Conduct the appropriate test, at 5% significance level, to test the hypothesis.

Solution:

1. State the hypothesis:

   Let p_A be the proportion of people in City A who are compliant with the norms and p_B be the proportion of people in City B who are compliant with the standards.

   Null hypothesis: H_0: $p_A = p_B$

   Alternate hypothesis: $H_1 : p_A ! = p_B$

   This would be a two-tail test, because the region of rejection could be located on either side.

2. Select the appropriate hypothesis test:

   - Number of samples: 2 (taking samples from two different cities)
   - Sample size: Large ($N_1 = 200$ and $N_2 = 230$)
   - What we are testing: Whether the proportion of passengers conforming with child safety guidelines is different for the two cities
   - Population characteristics: The distribution of the population is not known; population variances are unknown. Since sample sizes are large, we select the two-sample z-test for proportions

3. Fix the level of significance: α=0.05

4. Calculate the test statistic and p-value

   Using the following equation:

$$z = \frac{\left(\overline{p_1} - \overline{p_2}\right)}{\sqrt{\frac{p_c\left(1-p_c\right)}{N_1} + \frac{p_c\left(1-p_c\right)}{N_2}}}$$

   Calculate the p-value corresponding to this z-value using the *stats.norm.cdf* function:

   CODE:

```
x1,n1,x2,n2=110,200,106,230
p1=x1/n1
p2=x2/n2
pc=(x1+x2)/(n1+n2)
z=(p1-p2)/(((pc*(1-pc)/n1)+(pc*(1-pc)/n2))**0.5)
p=2*(1-stats.norm.cdf(z))
p
```

   Output:

```
0.06521749465064053
```

5. Comparing the p-value with the level of significance:

   Since the calculated p-value (0.065)>α(0.05), we fail to reject the null hypothesis.

6. Inference: There is no significant difference between the proportion of passengers in these cities complying with child safety norms, at a 5% significance level.

T-distribution

There may be situations where the standard deviation of the population is unknown, and the sample size is small. In such cases, we use the T-distribution. This distribution is also called Student's T distribution. The word "Student" does not assume its literal meaning here. William Sealy Gosset, who first published this distribution in 1908, used the pen name "Student," and thus this distribution became widely known as Student's T-distribution.

The following are the chief characteristics of the T-distribution:

- The T-distribution is similar in shape to a normal distribution, except that it is slightly flatter.
- The sample size is small, generally less than 30.
- The T-distribution uses the concept of degrees of freedom. The degrees of freedom are the number of observations in a statistical test that can be estimated independently. Let us understand the concept of degrees of freedom using the following example:

  Say we have three numbers: a, b, and c. We do not know their values, but we know the mean of the three numbers, which is 5. From this mean value, we calculate the sum of the three numbers - 15 (mean*number of values, 5*3).

  Can we assign any value to these three unknown numbers? No; only two of these three numbers can be assigned independently. Say we randomly assign the value 4 to a and 5 to b. Now, c can only be 6 since the total sum has to be 15. Hence, even though we have three numbers, only two are free to vary.
- As the sample size decreases, the degrees of freedom reduce, or in other words, the certainty with which the population parameter can be predicted from the sample parameter reduces.

  The degrees of freedom (*df*) in the T-distribution is the number of samples (n) -1, or in other words, $df = n - 1$.

The formula for the critical test statistic in a one-sample t-test is given by the following equation:

$$\mathbf{t} = \frac{\overline{x} - \mu}{s / \sqrt{n}}$$

where $\bar{x}$ is the sample mean, μ is the population mean, s is the sample standard deviation and n is the sample size.

One sample t-test

A one-sample t-test is similar to a one-sample z-test, with the following differences:

1. The size of the sample is small (<30).
2. The population standard deviation is not known; we use the sample standard deviation(s) to calculate the standard error.
3. The critical statistic here is the t-statistic, given by the following formula:

$$t = \frac{(\bar{x} - \mu)}{s / \sqrt{n}}$$

Two-sample t-test

A two-sample t-test is used when we take samples from two populations, where both the sample sizes are less than 30, and both the population standard deviations are unknown.

Formula:

$$t = \frac{\bar{x}_1 - \bar{x}_2}{\sqrt{S_p^2\left(\frac{1}{n_1} + \frac{1}{n_2}\right)}}$$

Where $\bar{x}_1$ and $\bar{x}_2$ are the sample means

The degrees of freedom: df=$n_1 + n_2 - 2$

The pooled variance: $S_p^2 = \frac{(n_1 - 1)S_1^2 + (n_2 - 1)S_2^2}{n_1 + n_2 - 2}$

Two-sample t-test for paired samples

This test is used to compare population means from samples that are dependent on each other, that is, sample values are measured twice using the same test group.

This equation gives the critical value of the test statistic for a paired two-sample t-test:

$$t = \frac{\bar{d}}{s / \sqrt{n}}$$

Where $\bar{d}$ is the average of the difference between the elements of the two samples. Both the samples have the same size, n.

S = standard deviation of the differences between the elements of the two samples =

$$\sqrt{\frac{\Sigma d^2 - (\Sigma d)^2 / n}{n-1}}$$

Solved examples: Conducting t-tests using Scipy functions

The Scipy library has various functions for the t-test. In the following examples, we look at the functions for the one-sample t-test, the two-sample t-test, and the paired t-test.

1. **One-sample t-test with Scipy**:

   Question: A coaching institute, preparing students for an exam, has 200 students, and the average score of the students in the practice tests is 80. It takes a sample of nine students and records their scores; it seems that the average score has now increased. These are the scores of these ten students: 80, 87, 80, 75, 79, 78, 89, 84, 88.

   Conduct a hypothesis test at a 5% significance level to verify if there is a significant increase in the average score.

Solution:

We use the one-sample t-test since the sample size is small, and the population standard deviation is not known. Let us formulate the null and alternate hypotheses.

$H_0{:}\mu = 80$
$H_1{:}\ \mu > 80$

First, create a NumPy array with the sample observations :

CODE:

```
a=np.array([80,87,80,75,79,78,89,84,88])
```

Now, call the *stats.ttest_1samp* function and pass this array and the population mean. This function returns the t-statistic and the p-value.

CODE:

```
stats.ttest_1samp(a,80)
```

Output:

```
Ttest_1sampResult(statist
ic=1.348399724926488,  pvalue=0.21445866072113726)
```

Decision: Since the p-value is greater than 0.05, we fail to reject the null hypothesis. Hence, we cannot conclude that the average score of students has changed.

2. **Two-sample t-test (independent samples)**:

Question: A coaching institute has centers in two different cities. It takes a sample of ten students from each center and records their scores, which are as follows:

Center A: 80, 87, 80, 75, 79, 78, 89, 84, 88

Center B: 81, 74, 70, 73, 76, 73, 81, 82, 84

Conduct a hypothesis test at a 5% significance level, and verify if there a significant difference in the average scores of the students in these two centers.

Solution:

We use the two-sample t-test since we are taking samples from two independent groups. The sample size is small, and the standard deviations of the populations are not known. Let the average scores of students in each of these centers be μ_1 *and* μ_2. The null and alternate hypothesis is as follows:

$$H_0{:}\mu_1 = \mu_2$$

$$H_1{:}\mu_1 ! = \mu_2$$

Create NumPy arrays for each of these samples:

CODE:

```
a=np.array([80,87,80,75,79,78,89,84,88])
b=np.array([81,74,70,73,76,73,81,82,84])
```

Call the *stats.ttest_ind* function to conduct the two-sample t-test and pass these arrays as arguments:

CODE:

```
stats.ttest_ind(a,b)
```

Output:

```
Ttest_indResult(statistic=2.1892354788555664,
pvalue=0.04374951024120649)
```

Inference: We can conclude that there is a significant difference in the average scores of students in the two centers of the coaching institute since the p-value is less than 0.05.

3. **T-test for paired samples**:

   Question: The coaching institute is conducting a special program to improve the performance of the students. The scores of the same set of students are compared before and after the special program. Conduct a hypothesis test at a 5% significance level to verify if the scores have improved because of this program.

Solution:

CODE:

```
a=np.array([80,87,80,75,79,78,89,84,88])
b=np.array([81,89,83,81,79,82,90,82,90])
```

Call the *stats.ttest_rel* function to conduct the two-sample t-test and pass these arrays as arguments:

CODE:

```
stats.ttest_rel(a,b)
```

Output:

```
Ttest_relResult(statistic=-2.4473735525455615,
pvalue=0.040100656419513776)
```

We can conclude, at a 5% significance level, that the average score has improved after the special program was conducted since the p-value is less than 0.05.

ANOVA

ANOVA is a method used to compare the means of more than two populations. So far, we have considered only a single population or at the most two populations. The statistical distribution used in ANOVA is the F distribution, whose characteristics are as follows:

1. The F-distribution has a single tail (toward the right) and contains only positive values, as shown in Figure 9-11.

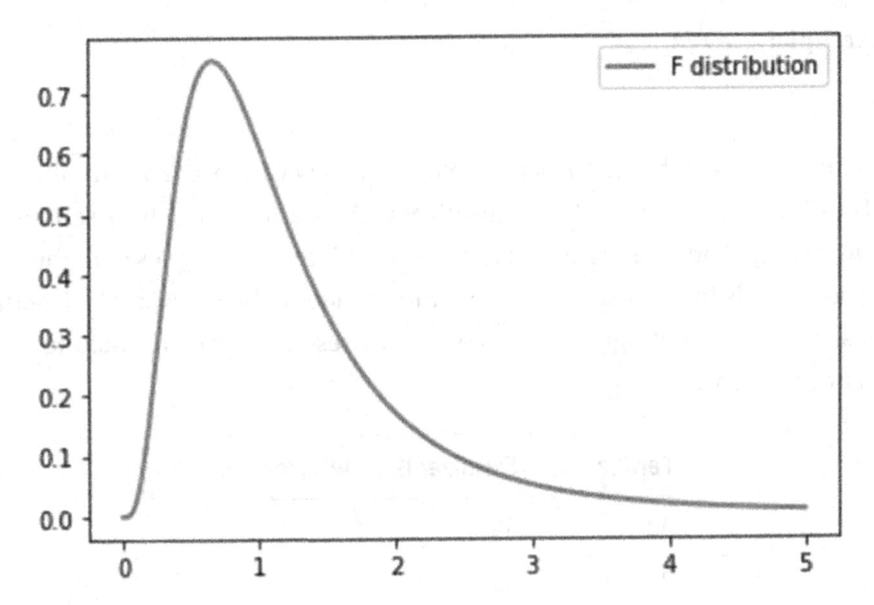

Figure 9-11. *Shape of the F-distribution*

2. The F-statistic, which is the critical statistic in ANOVA, is the ratio of variation between the sample means to the variation within the sample. The formula is as follows.

$$F = \frac{variation\ between\ sample\ means}{(variation\ within\ the\ samples)}$$

3. The different populations are referred to as treatments.

4. A high value of the F statistic implies that the variation between samples is considerable compared to variation within the samples. In other words, the populations or treatments from which the samples are drawn are actually different from one another.

5. Random variations between treatments are more likely to occur when the variation within the sample is considerable.

Solved example: ANOVA

Question:

A few agricultural research scientists have planted a new variety of cotton called "AB cotton." They have used three different fertilizers – A, B, and C – for three separate plots of this variety. The researchers want to find out if the yield varies with the type of fertilizer used. Yields in bushels per acre are mentioned in the below table. Conduct an ANOVA test at a 5% level of significance to see if the researchers can conclude that there is a difference in yields.

Fertilizer A	Fertilizer B	Fertilizer C
40	45	55
30	35	40
35	55	30
45	25	20

Solution:

1. State the null and alternative hypothesis:

   Let the average yields of the three populations be $\mu_1, \mu_2\ and \mu_3$

   Null hypothesis: $H_0 : \mu_1 = \mu_2 = \mu_3$

   Alternative hypothesis: $H_1 : \mu_1 ! = \mu_2 ! = \mu_3$

2. Select the appropriate test:

   We select the ANOVA test because we are comparing averages from three populations

3. Fix the level of significance: α=0.05

4. Calculate the critical test statistic/p-value:

   The *f_oneway* function gives us the test statistic or the p-value for the ANOVA test. The arguments to this function include three lists containing sample values of each of the groups.

CODE:

```
import scipy.stats as stats
a=[40,30,35,45]
b=[45,35,55,25]
c=[55,40,30,20]
stats.f_oneway(a,b,c)
```

Output:

```
F_onewayResult(statistic=0.10144927536231883,
pvalue=0.9045455407589628)
```

5. Since the calculated p-value (0.904)>0.05, we fail to reject the null hypothesis.
6. Inference: There is no significant difference between the three treatments, at a 5% significance level.

Chi-square test of association

The chi-square test is a nonparametric test for testing the association between two variables. A non-parametric test is one that does not make any assumption about the distribution of the population from which the sample is drawn. Parametric tests (which include z-tests, t-tests, ANOVA) make assumptions about the distribution/shape of the population from which the sample is drawn, assuming that the population is normally distributed. The following are some of the characteristics of the chi-square test.

- The chi-square test of association is used to test if the frequency of occurrence of one categorical variable is significantly associated with that of another categorical variable.
- The chi-square test statistic is given by: $X^2 = \frac{\Sigma(f_0 - f_e)^2}{f_e}$, where

  f_0 denotes the observed frequencies, f_e denotes the expected frequencies, and X is the test statistic. Using the chi-square test of association, we can assess if the differences between the frequencies are statistically significant.

- A contingency table is a table with frequencies of the variable listed under separate columns. The formula for the degrees of freedom in the chi-square test is given by:

  $df=(r-1)*)(c-1)$, where df is the number of degrees of freedom, r is the number of rows in the contingency table, and c is the number of columns in the contingency table.

- The chi-square test compares the observed values of a set of variables with their expected values. It determines if the differences between the observed values and expected values are due to random chance (like a sampling error), or if these differences are statistically significant. If there are only small differences between the observed and expected values, it may be due to an error in sampling. If there are substantial differences between the two, it may indicate an association between the variables.

- The shape of the chi-square distribution for different values of k (degrees of freedom) is shown in Figure 9-12. The chi-square distribution's shape varies with the degrees of freedom (denoted by k in Figure 9-12). When the degrees of freedom are few, it looks like an F-distribution. It has only one tail (toward the right). As the degrees of freedom increase, it looks like a normal curve. Also, the increase in the degrees of freedom indicates that the difference between the observed values and expected values could be meaningful and not just due to a sampling error.

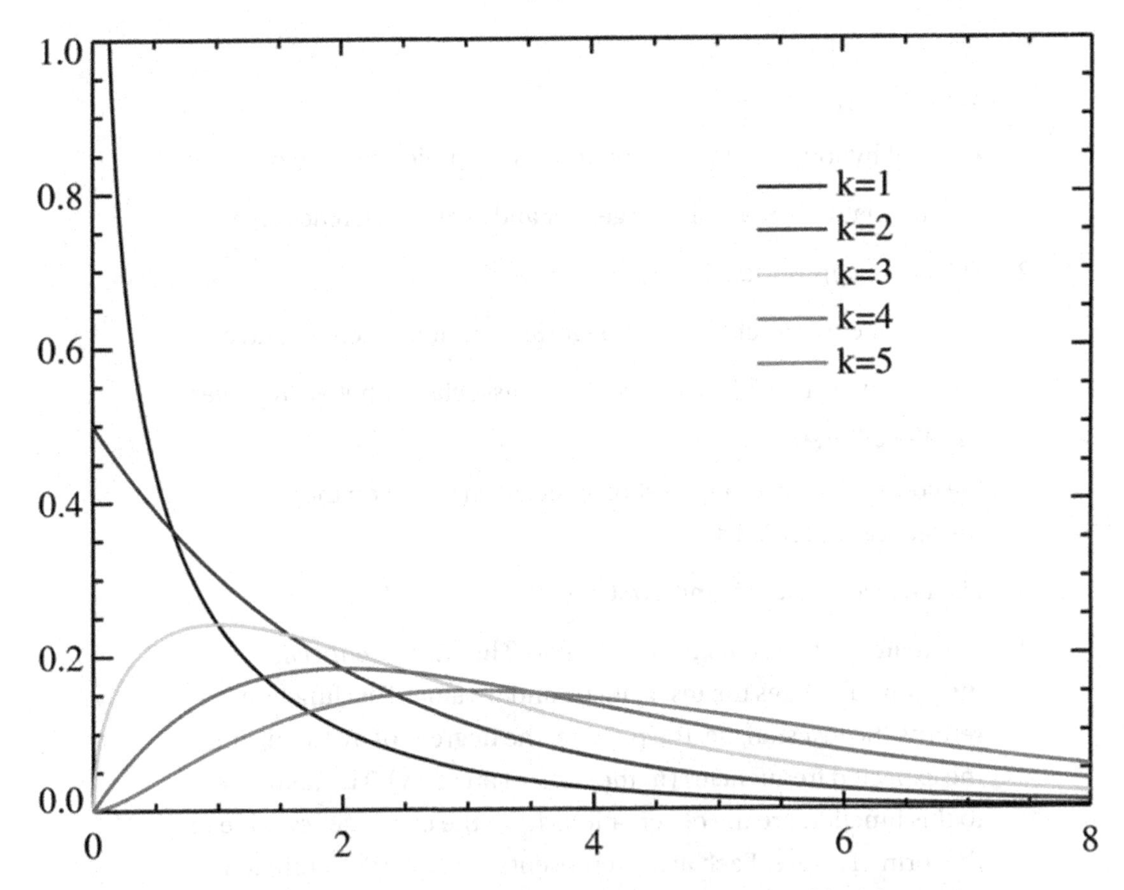

Figure 9-12. Chi-square distribution for different degrees of freedom

Solved example: Chi-square test

Question: A career counseling service guides students to help them understand their strengths and weaknesses so that they make appropriate career choices. They would like to assess if there is an association between the gender of a student and the career he or she chooses. The following table shows the number of males and females, and the careers (given by career IDs like I001, I002, etc.) they choose to pursue.

Career	Males	Females	Total
I001	41	79	120
I002	32	28	60
I003	58	78	130
I004	59	31	90

Answer:

1. State the hypothesis:
   - Null hypothesis: H_0: gender and career preference are not related
   - Alternative hypothesis: H_1: gender and career preference are related
2. Select the appropriate hypothesis test:
   - Number of variables: two categorical variables (gender and career)
   - What we are testing: Testing for an association between career and gender

   We conduct a chi-square test of association based on the preceding characteristics.
3. Fix the level of significance: α=0.05
4. Calculate the test statistic and p-value. The *chi2_contingency* function calculates the test statistic and p-value. This function returns the test statistic, the p-value, the degrees of freedom, and the expected frequencies (in the form of an array). The arguments to this function are the observations from the contingency table in the form of arrays. Each array represents a row in the contingency table.

CODE:

```
import scipy.stats as stats
observations=np.array([[41,79],[32,28],[52,78],[59,31]])
chi2stat,pval,dof,expvalue=stats.chi2_contingency(observations)
print(chi2stat,pval,dof,expvalue)
```

Output:

```
23.803453211665776 2.7454871071500803e-05 3 [[55.2 64.8]
 [27.6 32.4]
 [59.8 70.2]
 [41.4 48.6]]
```

The highlighted value in the preceding output is the p-value for this test.

5. Comparing the p-value with the level of significance:

   Since the calculated p-value (0.000027)<α(0.05), we reject the null hypothesis.

6. Inference: There is a significant association between the gender of the student and career choice, at a 5% significance level.

Caveat while using p-value:

The power of a hypothesis test is measured by its ability to yield statistically significant results, which is represented by a p-value that is less than 0.05. The results of many research trials and experiments conducted in the fields of medical and social sciences are presented using p-values. The p-value, however, is hard to interpret. It is also dependent on the sample size and the size of the bias that we measure. A result that is statistically significant does not conclusively disprove the null hypothesis or prove the alternate hypothesis. Confidence intervals are generally preferable to using p-values, as they are more easily interpretable.

Summary

1. Combinations refer to the number of ways in which we can select items, whereas permutations refer to the number of ways in which we can arrange them.

2. Probability is the likelihood of an event.

   Two events are independent when the probability of occurrence of one event does not affect the other. Independent events follow the special rule of multiplication, where $P(A \cap B)= P(A)*P(B)$.

   Mutually exclusive events are those that cannot occur together, and such events follow the special rule of addition, where $P(AUB)=P(A)+P(B)$.

3. The Bayes theorem calculates the posterior probability of an event, or in other words, the probability of a hypothesis being true given some evidence related to it.

4. A random variable takes the values associated with the outcomes of an experiment. There are two types of random variables: discrete (which take only a few values) and continuous (which can take any number of values).

5. Discrete variables can be used for binomial distributions (a single experiment repeated multiple times with each trial having two possible outcomes) or Poisson distributions (which model the number of occurrences that occur over an interval, given the average rate of occurrence).

6. The normal distribution is a symmetric bell-shaped curve, using a continuous random variable with most of its values centered around the mean. The standard normal distribution has a mean of 0 and a standard deviation of 1. The formula used in standard normal distributions is as follows:

$$z = \frac{(x - \mu)}{\sigma}.$$

7. Continuous distributions have various measures of central tendency (mean, median, mode), dispersion (range, variance, standard deviation), and shape (skewness is a measure of asymmetry while kurtosis is a measure of the curvedness of the distribution).

8. A sample is used when it is impractical to collect data about all the subjects in a large population. The main methods of collecting a sample are probability sampling (where subjects are randomly selected from a large population) and non-probability sampling (when data is not readily available, and samples are taken based on availability or access).

9. A hypothesis test is used to make an inference about a population based on a sample, but it does not conclusively establish anything about the population. It is only suggestive. The two kinds of estimates that can be made about a population from a sample are point estimates (using a single value) and interval estimates (using

a range of values). The confidence interval, which is the range of values within which the population mean lies, is an example of an interval estimate.

10. The null hypothesis indicates that nothing has changed, while the alternate hypothesis is used when we have reason to reject the null hypothesis. A Type 1 error occurs when the null hypothesis is incorrectly rejected when it is true. In contrast, a Type 2 error occurs when we fail to reject the null hypothesis when it is not true.

11. Either the test statistic (which is different for every hypothesis test) or the p-value can be used to decide whether or not to reject the null hypothesis. The p-value measures the likelihood that the observed data occurred merely by chance.

12. A two-tail test is used when we are testing whether the population parameter(s) is not equal to a particular value. In contrast, a one-tail test is used when the population parameter(s) is either greater than or less than a particular value.

    A one-sample test is used when a single sample is taken from a population, while a two-sample test compares samples taken from different populations.

13. Hypothesis tests can be either parametric (when we assume that the population from which a sample is drawn is normally distributed) or nonparametric (when we do not make such assumptions about the population distribution).

14. A parametric hypothesis test could be used to compare means using the z-test (when the sample size is large, and the population standard deviation is known) or the t-test (small sample size <30 and the population standard deviation is unknown). Z-tests can also be used to compare proportions. The ANOVA test is used when we need to compare the means of more than two populations. The chi-square test, one commonly used nonparametric test, is used for testing the association between variables.

Review Exercises

Question 1

Match the Scipy function (column on the right) with the appropriate hypothesis test (column on the left).

Hypothesis test	Scipy function
1. Chi-square	a. `stats.ttest_rel`
2. ANOVA	b. `stats.ttest_1samp`
3. Paired t-test	c. `stats.f_oneway`
4. One-sample t-test	d. `stats.chi2_contingency`
5. Two-sample (independent) t-test	e. `stats.ttest_ind`

Question 2

Skewness is a measure of:

1. Dispersion
2. Central tendency
3. Curvedness
4. Asymmetry

Question 3

Mr. J underwent a test for a widespread pandemic. The doctor made a clinical diagnosis that Mr. J does not have this illness. Later, when a blood test was conducted, it came out positive. Which of the following errors has the doctor committed?

1. Type 0 error
2. Type 1 error
3. Type 2 error
4. No error was committed

Question 4

Which of the following is correct?

1. A normal curve is an example of a mesokurtic curve
2. A playtykurtic curve is flat
3. A leptokurtic curve has a high peak
4. All of the above
5. None of the above

Question 5

Let us assume that you are testing the effectiveness of e-learning programs in improving the score of students. The average score of the students is measured before and after the introduction of the e-learning programs. After comparing the means using a hypothesis test, you obtain a p-value of 0.02. This means that

1. The probability of the null hypothesis being true is 2%.
2. You have definitively disproved the null hypothesis (which states that there is no difference between the average scores before and after the introduction of the e-learning programs).
3. There is a 2% probability of getting a result as extreme as or more extreme than what has been observed.
4. You have definitively proved the alternative hypothesis.

Question 6

A new health drink claims to have 100 calories. The company manufacturing conducts periodic quality control check by selecting random independent samples (100 calories). The most recent 13 samples of this drink show the following calorie values: 78, 110, 105, 72, 88, 107, 85, 92, 82, 92, 91, 82, 103. At a significance level of 5%, conduct a hypothesis test whether there is a change in the calorific value of the health drink from what was originally claimed.

Question 7

Silver Gym is offering a fitness–cum–weight loss program for its clients and claims that this program will result in a minimum weight loss of 3 kgs after 30 days. To verify this claim, 20 clients who joined this program were studied. Their weights were compared before and after they underwent this program.

The weights of the 20 clients before and after the fitness program are as follows:

before_weights=[56,95,78,67,59,81,60,56,70,78,84,71,90,101,54,60]

after_weights=[52,91,77,65,54,78,54,55,65,76,82,66,88,94,53,55]

Conduct the appropriate test to test the hypothesis that there is a 3-kg weight loss (assuming that the weights of the population are normally distributed).

Answers

Question 1

1-d; 2-c; 3-a; 4-b; 5-e

Question 2

Option 4: Asymmetry (skewness is a measure of asymmetry)

Question 3

Option 3: Type 2 error

A Type 2 error is committed when the null hypothesis is not rejected when it does not hold true. Here, the null hypothesis is that the patient does not have this illness. The doctor should have rejected the null hypothesis and made a diagnosis for this illness since the blood test result is positive.

Question 4

Option 4: All of the above

Question 5

Option 3

Remember that the p-value only gives us the probability of getting a result as extreme as or more extreme than what is observed. It does not prove or disprove any hypothesis.

Question 6

1. State the hypothesis:

   Let the mean calorie value of this drink be μ

   Null hypothesis: H_0: μ=100

   Alternative hypothesis: $H_1 : \mu != 100$

   This is a two-tail test.

2. Select the appropriate hypothesis test:

   We select the one-sample t-test based on the following characteristics:

   - Number of samples: one sample
   - Sample size: small (n=13)
   - What we are testing: mean calorific value
   - Population characteristics: population is normally distributed, and the population standard deviation is not known

3. Fix the level of significance: α=0.05

4. Calculate the test statistic and p-value:

   CODE:

   ```
   import numpy as np
   import scipy.stats as stats
   values=np.array([78,110,105,72,88,107,85,92,82,92,91,82,103])
   stats.ttest_1samp(values,100)
   ```

   Output:

   ```
   Ttest_1sampResult(statistic=-2.6371941582527527,
   pvalue=0.02168579243588164)
   ```

5. Comparison: Since the calculated p-value $<\alpha$, we reject the null hypothesis.

6. Inference: It can be concluded, at a 5% level, that there is a significant difference between the calorific value of the sample and that of the population.

Question 7

1. State the hypothesis:

   Let μ_d be the average difference in weights before and after the weight loss program for the population

   Null hypothesis: H_0:$\mu_d < 3$

   Alternative hypothesis: $\mu_d \geq 3$

   One-tail test since there is a greater-than-or-equal-to sign in the alternative hypothesis

2. Select the appropriate hypothesis test:

   - Number of samples: Two samples (taking two different samples with the same subjects)
   - Sample size: Small (20)
   - What we are testing: Testing the average difference in weight loss
   - Population characteristics: Distribution of population is normal, but population variances are not known.

     Based on the preceding characteristics and since the samples are related to each other (considering that we are comparing the weights of the same clients), we conduct a paired two-sample t-test.

3. Fix the level of significance: α=0.05

4. Calculate the p-value:

   The p-value can be calculated using the *stats.ttest_rel* equation as shown in the following code.

CODE:

```
import scipy.stats as stats
before_weights=[56,95,78,67,59,81,60,56,70,78,84,71,90,101,54,60]
after_weights=[52,91,77,65,54,78,54,55,65,76,82,66,88,94,53,55]
stats.ttest_rel(before_weights,after_weights)
```

Output:

```
Ttest_relResult(statistic=7.120275558034701,
pvalue=3.504936069662947e-06)
```

5. Conclusion/interpretation

Since the calculated p-value $<\alpha(0.05)$, we reject the null hypothesis.

It can be concluded that there is a significant difference between the two groups before and after the weight loss program.

Bibliography

https://upload.wikimedia.org/wikipedia/commons/5/5b/Binomialverteilung2.png

https://upload.wikimedia.org/wikipedia/commons/thumb/c/c1/Poisson_distribution_PMF.png/1200px-Poisson_distribution_PMF.png

https://upload.wikimedia.org/wikipedia/commons/thumb/f/fb/Normal_distribution_pdf.svg/900px-Normal_distribution_pdf.svg.png

https://commons.wikimedia.org/wiki/File:Standard_deviation_diagram.svg

https://upload.wikimedia.org/wikipedia/commons/d/d8/Normal_distribution_curve_with_lower_tail_shaded.jpg

https://upload.wikimedia.org/wikipedia/commons/thumb/c/cc/Relationship_between_mean_and_median_under_different_skewness.png/1200px

https://commons.wikimedia.org/wiki/File:Kurtosimailak_euskaraz.pdf

https://upload.wikimedia.org/wikipedia/commons/2/2d/Empirical_CLT_-_Figure_-_040711.jpg

https://upload.wikimedia.org/wikipedia/commons/1/10/Region_of_rejections_or_acceptance.png

Wikimedia Commons (https://commons.wikimedia.org/wiki/File:Chi-square_distributionPDF.png)

Index

G. Rajagopalan, *A Python Data Analyst's Toolkit*, https://doi.org/10.1007/978-1-4842-6399-0

D

E

F

G

H

I

J

O

P

T, U

V

W, X, Y

Z

Printed in the United States
By Bookmasters